Mechanics of Materials Laboratory Course

Synthesis SEM Lectures on Experimental Mechanics

Editor
Kristin Zimmerman, *SEM*

Synthesis SEM Lectures on Experimental Mechanics follow the technical divisions and the direction of The Society for Experimental Mechanics (SEM). The SEM is composed of international members of academia, government, and industry who are committed to interdisciplinary application, research and development, education and active promotion of experimental methods to: (a) increase the knowledge of physical phenomena; (b) further the understanding of the behavior of materials, structures and systems; and (c) provide the necessary physical basis and verification for analytical and computational approaches to the development of engineering solutions. The members of SEM encompass a unique group of experimentalists, development engineers, design engineers, test engineers and technicians, and research and development scientists from industry and educational institutions working in materials; modeling and analysis; strain measurement and structural testing.

Mechanics of Materials Laboratory Course
Ghatu Subhash and Shannon Ridgeway
2018

The Old and New... *A Narrative on the History of the Society for Experimental Mechanics*
Cesar A. Sciammarella and Kristin B. Zimmerman
2018

Hole-Drilling Method for Measuring Residual Stresses
Gary S. Schajer and Philip S. Whitehead
2018

Mechanics of Materials Laboratory Course

Ghatu Subhash and Shannon Ridgeway

ISBN: 978-3-031-79720-0   paperback
ISBN: 978-3-031-79721-7   ebook
ISBN: 978-3-031-79722-4   hardcover

DOI 10.1007/978-3-031-79721-7

A Publication in the Springer series
SYNTHESIS SEM LECTURES ON EXPERIMENTAL MECHANICS

Lecture #2
Series Editor: Kristin Zimmerman, *SEM*
Series ISSN
Print 2577-6053   Electronic 2577-6088

Mechanics of Materials Laboratory Course

Ghatu Subhash and Shannon Ridgeway
University of Florida

SYNTHESIS SEM LECTURES ON EXPERIMENTAL MECHANICS #2

ABSTRACT

This book is designed to provide lecture notes (theory) and experimental design of major concepts typically taught in most Mechanics of Materials courses in a sophomore- or junior-level Mechanical or Civil Engineering curriculum. Several essential concepts that engineers encounter in practice, such as statistical data treatment, uncertainty analysis, and Monte Carlo simulations, are incorporated into the experiments where applicable, and will become integral to each laboratory assignment. Use of common strain (stress) measurement techniques, such as strain gages, are emphasized. Application of basic electrical circuits, such as Wheatstone bridge for strain measurement, and use of load cells, accelerometers, etc., are employed in experiments. Stress analysis under commonly applied loads such as axial loading (compression and tension), shear loading, flexural loading (cantilever and four-point bending), impact loading, adhesive strength, creep, etc., are covered. LabVIEW software with relevant data acquisition (DAQ) system is used for all experiments. Two final projects each spanning 2–3 weeks are included: (i) flexural loading with stress intensity factor determination and (ii) dynamic stress wave propagation in a slender rod and determination of the stress–strain curves at high strain rates.

The book provides theoretical concepts that are pertinent to each laboratory experiment and prelab assignment that a student should complete to prepare for the laboratory. Instructions for securing off-the-shelf components to design each experiment and their assembly (with figures) are provided. Calibration procedure is emphasized whenever students assemble components or design experiments. Detailed instructions for conducting experiments and table format for data gathering are provided. Each lab assignment has a set of questions to be answered upon completion of experiment and data analysis. Lecture notes provide detailed instructions on how to use LabVIEW software for data gathering during the experiment and conduct data analysis.

KEYWORDS

errors and uncertainty propagation, stress and strain measurement, calibration, Monte Carlo uncertainty estimation, axial, flexural, creep and impact loading, adhesive bonding, stress concentration, wave propagation, impact

In theory, theory and practice are the same.
In practice, they are not.

… Albert Einstein

*We dedicate this book to our parents and families
whose sacrifices and support have shaped our lifes.*

… Subhash and Shannon

Contents

Preface

This book is an effort to bring together theory and experiments under one roof to facilitate student learning in the Mechanics of Materials Laboratory course. Although students may have already taken the theory course on Mechanics of Materials (Strength of Materials), it has been our observation that many of them struggle to put together concepts from several theory chapters to effectively discuss relevant results and observations from one experiment. Therefore, we have made this book self-sufficient in its contents so that students (and instructors) do not have to go back to their textbooks. The experiments have been chosen to expose students not only to the basic concepts of Mechanics of Materials (MOM) course but also to several essential concepts that engineers encounter in practice such as statistical data treatment and uncertainty that always accompanies experimentally derived data. The concepts have also been extended to include uncertainty propagation and use of Monte Carlo simulations to estimate variability in a measured quantity. All these concepts have been made integral to each laboratory assignment as and when applicable.

The overall organization of this book can be divided into three parts: (i) statistical treatment of experimental data, use of LabView, and uncertainty analysis; (ii) experiments covering basic concepts of mechanics of materials; and (iii) final projects that are open-ended laboratory experiments with advanced concepts. Each experiment in the first two sets can be completed in one lab session (equivalent to two lecture hours) whereas, the third set will require at least two laboratory sessions by a group of 2–4 students, each student contributing to different components of the laboratory report. In each semester, it is recommended that an instructor choose the experiment from set (i), three to four experiments from set (ii), and one of the two final projects from set (iii). Lab 1 also provides additional details that are relevant to all other labs such as use of data acquisition (DAQ) module, laboratory report writing guidelines, and basics of LabVIEW programming instructions.

This laboratory course emphasizes commonly used strain (stress) measurement techniques such as strain gages in conjunction with Wheatstone bridge. Load cells, LVDT, and accelerometers will also be used. Basic concepts such as stress analysis under axial loading (compression and tension), biaxial loading, flexural loading, impact loading, adhesive strength, creep, etc. are covered. LabVIEW software with a relevant data acquisition (DAQ) system is recommended in all experiments. Two final projects encompassing flexural loading with stress intensity factor determination, and stress wave propagation and determination of high strain rate stress–strain response of materials are included.

Each lab is organized into two parts. Part A includes theoretical concepts that are pertinent to each experiment. When necessary, example problems and questions for further in-depth discussion are included. Part B contains laboratory component with relevant objectives, equipment needs, instructions for conducting experiment, and discussion points to be included in the report. Prelab assignments are included to prepare students for each lab. This assignment may include questions on theoretical concepts, limitations of the equipment and sensors, calculations needed for the lab, etc. At the end of each laboratory exercise, instructions or website addresses for securing off-the-shelf items needed for the experiment are provided. Instructors are encouraged to follow the designs shown for each equipment and build their own or modify as they see fit. A brief summary of each laboratory exercise is given below.

Lab 1 deals with *statistical analysis* of data gathered from DAQ, treatment of noise, and uncertainty quantification and propagation. In Lab 2, students will design and build a *load transducer* using a bonded a strain gage on a cantilever beam. They will calibrate the transducer, measure the weight of a beverage can, discuss source of errors, and quantify uncertainty in measurements. Use of resistance strain gages and Wheatstone bridge are introduced.

In Lab 3, uniaxial *tensile and compression testing* will be conducted on several types of materials. Both load-controlled and displacement-controlled stress–strain will be derived. Tension testing will be conducted on a metal, a polymer, and a uniaxial fiber reinforced composite. Students will not be told the exact type of metal being tested but based on the stress–strain curve (modulus, yield strength, and ultimate strength), they will have to determine the type of metal and its heat-treatment conditions. In addition to discussing various feature in the stress–strain curve, they will also discuss uncertainty in data and errors in the measurements. Monte Carlo method is introduced to estimate the error and uncertainty in the modulus measurement. For load-based stress–strain response, a thin copper wire is anchored to a LVDT with a carriage to support weights. For a displacement-based stress–strain curve a Material Test System (MTS) is used. Finally, compression testing using a cylindrical ceramic (Plaster of Paris) specimen is introduced. The failure plane in each loading scenario and material type is discussed using Mohr's circle. Lab 4 involves measuring pressure inside a *pressure vessel* and determination of failure plane under a biaxial stress state. Student will bond a 3-gage strain rosette at an arbitrary angle on the surface of a soda can and determine the internal pressure with which the contents are sealed. Students will utilize, strain transformations, stress–strain relationships, principal stresses, pressure vessel theory, and Mohr's circle to determine the internal pressure and failure plane. They will explore how to determine pressure with only two strain gages and eventually with only one gage. Errors in the measurements and uncertainty in the measured pressure value are estimated. Lab 5 deals with *strength of adhesive joints*. Students will bond two pieces of metal with different adhesives and evaluate their bond strength. They will learn to differentiate between adhesive failure and cohesive failure, and single-shear and double-shear loading. Uncertainty in the measured values and how do errors propagate in estimation of shear strength are explored.

In Lab 6, creep testing of a metallic (copper) wire is undertaken. The short-term creep behavior is determined by applying constant load and measuring elongation over time. Agreement with conventional creep models is examined. In Lab 7, *impact testing using Charpy tester* is explored using a notched metallic specimen and a brittle material. The impact resistance is determined in terms of energy absorbed during the fracture of these materials. Both physics-based approach (kinetic energy difference) and mechanics-based approach will be used to calculate the impact resistance and energy absorbed during the impact.

The last two labs are *final projects* which are open ended. In addition to experiments, students can utilize finite element software packages (e.g., ABAQUS, Solidworks, or develop their own finite element code) to simulate the experiment, validate their experimental results, and gain in-depth understanding of the stress or strain distribution in the test material. Accordingly, Lab 8 is designed to provide a comprehensive understanding of *flexural loading*. Using a custom-built four-point simply supported rectangular beam with holes and notches and bonded strain gages at critical locations, students will measure stress distribution and beam deflections. They will calculate stress concentration factors near notches and holes. As an open-ended project, students may use digital image correlation (DIC) method or photoelastic coatings to measure strains or stresses at critical locations, determine stress intensity factors, and compare these values to those calculated from MOM theory. They are encouraged to use commercial software packages and simulate the experiment and compare the stress distribution and stress concentration factors with those measured in the experiments and calculated from theory. Finally, in Lab 9 students get familiarity with the theory of *1-D stress wave propagation* in long elastic rods with strain gages bonded on the surface at mid-way point along the length. They will investigate the wave propagation behavior and reflection and transmission behaviors at the free and fixed ends of the rods. They will compare theoretical wave velocity with the experimentally measured wave velocity and identify the reasons for discrepancies. In the next step, students will use this knowledge to determine the dynamic stress–strain behavior of a metal and a ceramic using a split Hopkinson pressure bar, and compare this behavior to the qusistatic static stress–strain response. Students are encouraged to use ABAQUS® or DYNA explicit finite element codes to simulate the wave propagation in long rod and also to determine the stress–strain response of a metallic material using any of the strain-hardening models.

The authors' main intent of embarking on this book journey is to facilitate teaching Mechanics of Materials Laboratory Course by providing a structured instructional material and easy-to-follow laboratory experimental plan. In recent years, many universities have moved away from offering hands-on laboratory courses for a variety of reasons. Difficulties are also experienced by junior faculty in assembling a set of experiments that are easy to design and deliver to a large body of students. To this end, we hope that this book will ease the burden of putting together a laboratory course that will capture the basic elements of Mechanics of Materials.

Finally, the authors are thankful to many previous laboratory instructors and teaching assistants who have made numerous contributions and improvements to the Mechanics of Materials Laboratory at the University of Florida. The authors are grateful to their initiative and sincere efforts for improvement. We dedicate this book to all those students who appreciate and enjoy their first hands-on laboratory experience in Mechanics of Materials.

Ghatu Subhash and Shannon Ridgeway
April 2018

LABORATORY 1

Dynamic Data Acquisition and Uncertainty in Measurements

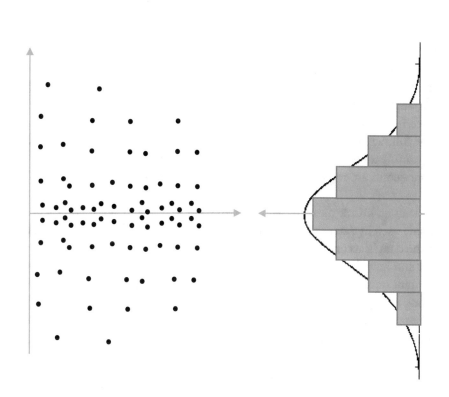

PART A: THEORY

1.1 STATISTICAL TREATMENT OF DATA AND UNCERTAINTY IN MEASUREMENTS

Engineering components and structures are made of materials that are inherently prone to variations in microstructure (grain size and grain distribution, defect size, and their distribution) and composition. Therefore, these components exhibit a distribution of properties and variability in performance even when the material is obtained from same batch or same process-run. The distribution of properties in materials leads to deviations or variability in component performance and product life. In addition, errors do occur when measurements are made by personnel, and the nature and distribution of errors may vary when different personnel or when different sensors or types of sensors are used to generate data. There are also many other unknowns including unexpected conditions in which a component is expected to perform (i.e., extreme weather or using your car in Alaska where conditions are extreme). These uncertainties in material properties and errors in measurements propagate from material stage to component level, and eventually to the final product performance index. In order to effectively deal with these issues in design of reliable engineering structures, it is essential that these uncertainties are quantified in a scientific manner. In the following, a brief discussion on types of errors, statistical treatment of data, and uncertainty quantification and propagation, as applied to this laboratory course, are briefly presented. If more details are required, the student is referred to relevant books at the end of this chapter.

While dealing with a measurement system, one must be aware of precise definition of terms commonly used in that context. For example, what is the difference between accuracy and precision? If the data is accurate, does it mean it is also precise? What is meant by sensitivity of an instrument and how does it relate to the minimum or maximum value that can be measured by that instrument or sensor? In sports, a shooter may be aiming to hit a bulls-eye or a baseball pitcher is aiming a target at the center, as shown in Fig. 1.1. This is the action (intention) but the result (the outcome) may or may not be the same every time. Then how do we define accuracy? Here are definitions of some relevant terms.

- *Accuracy:* It refers to the closeness of your action (measurement) to the desired value (true/actual/correct value or the anticipated result). In this course, it defines the closeness of a measured value to the actual or correct value.

- *Precision:* This term refers to repeatability of a measurement, i.e., the sensor reads the same value when the measurement is repeated several times. It has no relationship to accuracy.

Both these terms are illustrated with the following two examples: one numerical and the other pictorial.

- Let us say, the true dimension of a component is 1″ and the measurement shows 1.23452, then only the first digit is accurate. If repeated measurements show 1.23461, 1.23415, etc., then the measurements appear to be precise to 4 digits, but none of them are accurate.

- A baseball pitcher is throwing a ball toward the home plate as depicted in Fig. 1.1. In (a) the pitcher is accurate and precise where as in (b) he/she is only precise but not accurate (this pitcher has a bias towards to the left top corner). In (c) he/she is less accurate and not precise at all and in (d) he/she is neither.

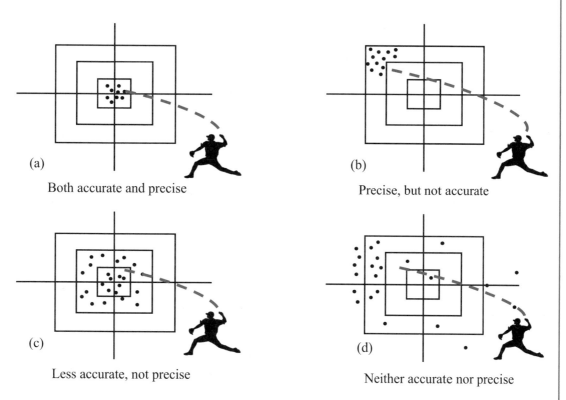

(a) Both accurate and precise

(b) Precise, but not accurate

(c) Less accurate, not precise

(d) Neither accurate nor precise

Figure 1.1: Illustration of concepts of accuracy and precision using a baseball pitcher.

- *Sensitivity:* Sensitivity of a measurement system or an instrument refers to the smallest measurement that can be made with that instrument. It often refers to the lowest gradation marker or the smallest digital reading the instrument is capable of. For example, in Fig. 1.2, the thermometer can read between 93°F and 101°F at 2°F increments. The minimum reading it can make is 93°F which is its sensitivity.

- *Resolution:* Refers to the minimum separation between two graduation marks or readings where they are still seen as two distinct values which can be measured. In Fig. 1.2, the resolution of the thermometer is 2°F.

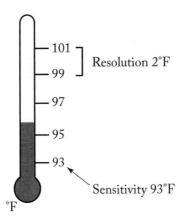

Figure 1.2: Illustration of sensitivity and resolution concepts using an analog thermometer.

1.2   STATISTICAL DATA REPRESENTATION OF INFINITE DATA

While making measurements using any electronic data acquisition system, the measured value will not be a constant single number over time, but changes slightly with time due either to noise in the system or to small fluctuations (thermal oscillations) in the electronics. Hence, the data falls over a range but sometimes may be centered around a fixed value, usually called "mean" which may also change over time. This random scatter can be represented by the Fig. 1.3, where the mean value is shown to be constant.

This example can be related to many kinds of data. For example, life expectancy in a population, miles driven by a particular model of an automobile when the first repair is needed, failure data on tensile test specimens, polling data on a specific issue from a population, etc.

Let N be the total number of sample measurements ($X_i, i = 1, 2, 3 \ldots N$) of a value X_{true}, such that N is large (i.e., $N \rightarrow \infty$). The deviation of a measured value (X_i) from the actual value (X_{true}) is the "random error" ε_i. These concepts are illustrated in Fig. 1.3. If the data is gathered over a period of time (or when sufficient number of data points are available), then the mean (μ) of the data can be written as:

$$\mu = Lim_{N \rightarrow \infty} \frac{1}{N} \sum_{i=1}^{\infty} X_i = \frac{1}{N} (X_1 + X_2 + \ldots X_N) \tag{1.1}$$

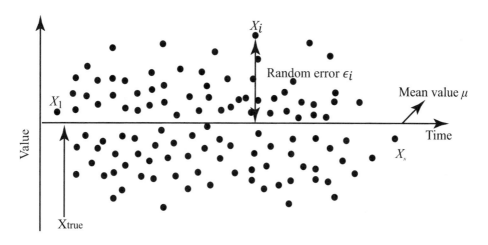

Figure 1.3: Representation of random error and mean values for a data set measured over time.

To understand the distribution of this data, we partition the data into different sets or intervals, with each interval spread over a certain range. The discretized data is plotted as shown in Fig. 1.4b, where x-axis represents *measured values* split into a set of ranges (intervals) and y-axis is the *number of data points* in a given interval (range). This data is presented as rectangles whose width is the range of values in a given interval and height is the number of values that fall within this range.

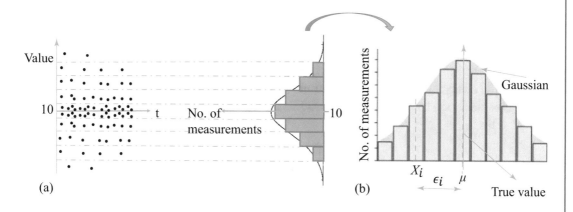

Figure 1.4: (a) Illustration of data partitioning to obtain data distribution and (b) approximation to Gaussian distribution and representation of random error and mean values.

For infinite data, this plot may be approximated by a Gaussian curve, as shown in Fig. 1.4b. The curve usually centers on the mean value where maximum number of data points (measurements) fall. The breadth of the curve represents the total range of values that were collected (minimum to the maximum value). The deviation from the mean value is the "random error."

In many instances, the system can have a constant bias, where majority of the data falls on one side of the true value. A bias (β) can be a persistent non-zero value in an instrument, e.g., initial non-zero reading on a "weighing machine" or a "bathroom scale." If we know the bias and depending on whether it is positive or negative, the final reading needs to be corrected either by subtracting (for positive value of bias) or adding (for negative value of bias) the bias value. All measurements are affected by bias and when it is present, the data centers on a new value. The total error (δ_i) for a single measurement is $\delta_i = \beta \pm \varepsilon_i$. The "bias error" ($\beta$) can be represented on a Gaussian plot, as shown in Fig. 1.5.

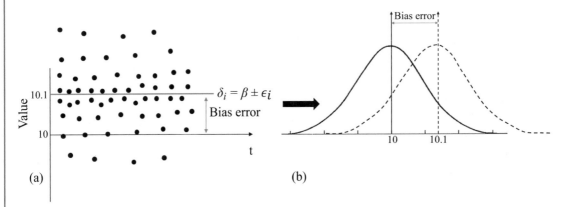

Figure 1.5: (a) Data distribution when bias exists in the system and (b) its Gaussian representation.

Standard Deviation (SD):   Instead of focusing on how much each measured data point deviates from the mean value, we calculate a number that measures *what percentage of data points* fall within a set-deviation from the mean value. This is represented by standard deviation (σ) and is calculated by

$$\sigma = \lim_{N \to \infty} \left[\frac{1}{N} \sum (X_i - \mu)^2 \right]^{1/2} \tag{1.2}$$

For a normal distribution, the percentage of data that falls within a given range of standard deviation from the mean value are shown in Table 1.1. On a Gaussian plot, σ is represented, as shown in Fig. 1.6. From Table 1.1, it is seen that 68.3% of data fall within 1σ from the mean, i.e., 68.3% of data fall within the $\mu \pm \sigma$ range.

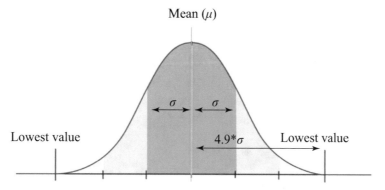

Figure 1.6: Representation of data, mean value, and standard deviation of a standard normal distribution.

Table 1.1: Relationship between percent data and standard deviation

Probability or CI or data fraction	Reading will fall within $\mu \pm k\sigma$ where k value is given by
50%	0.675
68.30%	1
95%	1.96
99%	2.58
99.9999%	4.9

Another interpretation of standard deviation "σ" is the probability that a measured data point falls within a certain range around the mean value for a normal distribution. For example, there is 68.3% probability that a measured data point falls within 1 standard deviation. Similarly, 50% of the data falls within 0.675σ from the mean, 95% of data falls within 1.96σ, and almost all the data (i.e., 99.9999%) falls within 4.9σ range from the mean. The percentage data can also be interpreted as confidence interval (CI). For example, for $1*\sigma$, there is 68.3% probability (or confidence) that, when you take infinite (large number) number of measurements, the next value will fall in this range (i.e., $\mu \pm \sigma$). Similarly, there is 50% probability (or 50% confidence) that the next measurement will fall within $\mu \pm 0.675*\sigma$ range.

Note that almost all the experiment data (i.e., 99.9999%) falls within $\pm 4.9*\sigma$ ($\sim 5\sigma$). However, in reality we rarely have infinite number of data points. Also, in industry, most of the experimental data is analyzed for 95% CI.

1.3   STATISTICAL DATA REPRESENTATION FOR FINITE DATA

When the sample population is finite ($N \ll \infty$), the average and SD are given by

$$\bar{X} = \frac{1}{N} \sum_{i=1}^{N} X_i \tag{1.3}$$

and

$$S_X = \text{approximate of standard deviation} = \left\{ \frac{1}{N-1} \sum_{i=1}^{N} (X_i - \bar{X})^2 \right\}^{1/2} \tag{1.4}$$

Although we often use this equation, it is important for student to realize that this equation underestimates the standard deviation when the data available (i.e., N) is small. For example, for $N = 10$ it is about 3% too small, and for $N = 4$ it is about 8% too small. This may be significant because in many practical applications (where it is too expensive to conduct repeated tests, e.g., NASA flight tests, airplane crash tests, etc.) one may have very few samples to calculate standard deviation.

One would wonder how is this approximate value (S_X) related to standard deviation (σ) of infinite measurements.

Let us assume that in a large class each lab group of 10 students performed 6 tensile tests. The average of these six tests is going to be slightly different for each group. So when one group reports their average value, what confidence level do we have in this number? How can we make this number representative of infinite measurements?

For this purpose, the uncertainty ($P_{\bar{X}}$) associated with this average or the uncertainty in the standard deviation of the average value should be defined. Here we define t-distribution such that this random uncertainty associated with the average for each group is defined as $P_{\bar{X}}$ so that we can define average of infinite data points as

$$\mu = \bar{X} + P_{\bar{X}} \quad \text{where} \quad P_{\bar{X}} \quad \text{is defined as} \quad P_{\bar{X}} = t S_{\bar{X}},$$

where $S_{\bar{X}}$ is the standard deviation of the limited data sample defined as

$$S_{\bar{X}} = \frac{S_X}{\sqrt{N}} \tag{1.5}$$

and t is a number related to how many times the measurement has been taken and at what confidence interval the value needs to be reported.

We will use 95% CI in this course because it is a common practice in industry. The values of t for 95% confidence interval are given in Table 1.2. Note that these values change with CI chosen.

Table 1.2: t-values in normal distribution for 95% confidence interval for finite number of tests

N, Number of Measurement	t for 95% Confident Interval
2	12.706
3	4.303
4	3.182
5	2.776
6	2.571
10	2.201
30	2.045
∞	1.96

Note that for infinite number of measurements at a CI of 95%, $t = 1.96$, the same value of k in Table 1.1. In general, for measurements $N > 10$, we use $t = 2.0$.

Example 1.1
Six cell phones $(X_{i=1,2,...,6})$ were used to estimate the mean failure time in hours with 95% confidence in Table 1.3. Find the mean for 95% CI.

Table 1.3: Cell phones used to estimate the mean failure time

Cell phone #	Failure time (hours)
X_1	1540
X_2	1390
X_3	1275
X_4	1464
X_5	1252
X_6	1314

Solution:

Average
$$\bar{X} = \frac{1}{N} \sum_{i=1}^{N} X_i = 1372 \text{ h}$$

Approx. Standard Deviation
$$S_X = \left\{ \frac{1}{N-1} \sum_{i=1}^{N} (X_i - \bar{X})^2 \right\}^{0.5} = 114 \text{ h}$$

From Table 1.2, for six tests, $t = 2.571$

Mean
$$\mu = \bar{X} \pm t \frac{S_X}{\sqrt{N}} = 1372 \pm 2.571 \left(\frac{114}{\sqrt{6}} \right) \text{ h}$$

$$\mu = 1372 \pm 120 \text{ h}$$

Example 1.2

A manufacturer tests seven cars to estimate the mean operation time without any repairs. Find the mean runtime of the vehicle lot so that they can provide a Guarantee Certificate at 95% confidence interval (Car number $= X_i$).

Table 1.4: Cars and days without repairs

Car	Days
X_1	1095
X_2	1110
X_3	1132
X_4	1064
X_5	1047
X_6	1156
X_7	1084

Solution:

Average
$$\bar{X} = \frac{1}{N} \sum_{i=1}^{N} X_i = 1098.3 \text{ days}$$

Standard Deviation
$$S_X = \left\{ \frac{1}{N-1} \sum_{i=1}^{N} (X_i - \bar{X})^2 \right\}^{0.5} = 37.9$$

From Table 1.2, for 7 tests, $t = 2.365$

$$\text{Mean} \quad \mu = \bar{X} \pm t \frac{S_X}{\sqrt{N}} = 1098.3 \pm 2.365 \left(\frac{37.9}{\sqrt{7}}\right) \text{ days}$$

$$\mu = 1098.3 \pm 33.87 \text{ days}$$

Note that for 7 cars tested, the standard deviation is 37.9 days but when extended for infinite measurements its deviation from mean value is 33.87 days for 95% confidence interval.

1.4 UNCERTAINTY ANALYSIS

As mentioned in the introductory paragraph of this laboratory, uncertainty can arise from many sources; batch to batch variation in materials or products, dimensional inaccuracies, or variabilities, use of product in unintended circumstances, etc. Uncertainty also arises from lack of knowledge on all possible sources of uncertainty. In this laboratory, we will only concentrate on two sources of uncertainty. They are:

1. Uncertainty inherent in the physical system, i.e., variations in material microstructure, component size, dimensional variability, imperfections in shape and surface finish, presence of joints, material properties, uncertainty arising from use in conditions other than for which it is designed, etc. Basically, we realize that each component/system or product is different from the other.

2. Uncertainty in the measurement system or sensors (random error, calibration error, etc.), operator variability, etc.

   For example, if one is conducting 10 tension tests, there will be scatter in the measured data. Each material (test specimen) is different because there are differences in the material microstructure (grain size, impurity content, distribution of porosity and phases, defects, etc.). Also, when making measurements additional uncertainty (scatter) arises from the measurement of dimensions, noise, and bias from the instruments, noise from sensors, etc.

   How does uncertainty propagate in the system and influence the final product performance? For example, when a car is manufactured, each component (e.g., engine, chassis, tires, transmission, etc.) brings its own uncertainty from different suppliers. Each component consists of many subparts (bolts, nuts, joints, etc.), and they have their own uncertainty (or variability). So, we must have a method to determine how uncertainty propagates from these small parts, to subparts, and finally to the system (car).

   The overall uncertainty (U_X) is given by

$$U_X^2 = \varepsilon_X^2 + \beta_X^2 \quad (3) \qquad \text{where} \begin{cases} \varepsilon_X = \text{Random uncertainty} \\ \beta_X = \text{Systematic (bias) uncertainty} \end{cases} \qquad (1.6)$$

For simplicity, assume that there is no systematic error or uncertainty, i.e., $\beta_X = 0$.

Let us assume that there are X_i variables ($i = 1, \ldots, n$) measured during an experiment and used to calculate a result r. Also assume that the result r is linearly dependent on the X_i (in the range of variation of the X_i's due to randomness), then the uncertainty in experimental result can be approximated by U_r, given by

$$U_r = \left[\sum_{i=1}^{n} \left(U_{X_i} \frac{\partial r}{\partial X_i} \right)^2 \right]^{1/2} \quad \text{where} \quad \begin{cases} r = \text{experimental results} \\ X_i = \text{measured variable} \end{cases} \tag{1.7}$$

or

$$U_r^2 = \left(\frac{\partial r}{\partial X_1} \right)^2 U_{X_1}^2 + \left(\frac{\partial r}{\partial X_2} \right)^2 U_{X_2}^2 + \ldots \left(\frac{\partial r}{\partial X_n} \right)^2 U_{X_n}^2 \tag{1.8}$$

where U_{X_i} is the uncertainty in the variable X_i and the partial derivative $\frac{\partial r}{\partial X_i}$ is the sensitivity of the results (r) to the variable X_i. This method of finding uncertainty will be referred to as "root sum squared" (RSS) method.

The term sensitivity can be explained as follows: For simplicity, let us assume that a car manufacturer is planning to provide a warranty (result) based on two components, engine and brakes (variables), it receives from two suppliers. In laboratory tests, 95% of engines successfully ran for 50,000 miles and a similar percentage of breaks for 75,000 miles. Here we can consider the 95% as CI. Obviously, from the perspective of the warranty for the car, there is a greater chance for engine to fail early than brakes. So the result (warranty or the service life) is more sensitive to the lifetime of the engine than brakes.

The uncertainty from each variable (X_i) is calculated as the product of the sensitivity $\left(\frac{\partial r}{\partial X_i} \right)$ of the result to a given variable and the variability in that variable U_{X_i}. This is represented by the term in the inner parenthesis of Eq. (1.7). These products for each variable are squared and summed to get the overall uncertainty square, as shown in Eqs. (1.7) and (1.8).

Example 1.3

During a year, number of students in a class is 50 ± 2, and each student reads 10 ± 1 books. What is the uncertainty (variability or scatter) in the total number of books read? (It is assumed that the mean of the result r is obtained by multiplying the mean of each of the variables X_i.)

Solution:

Number of students $= S$,   No. of books read by each student $= B$.

Total number of books read, $N = B.S$.

Here $\begin{cases} X_i = \text{Measured variables } (S \ \& \ B), \ X_1 = B = 10 \text{ and } X_2 = S = 50 \\ r = \text{experimental result} = N \\ U_{X_i} = \text{uncertainty in the measured variable} \\ U_B = \pm 1 \text{ book} \\ U_S = \pm 2 \text{ students} \\ U_r = \text{uncertainty in the result (total number of books read)} \end{cases}$

$$U_r = \left[\sum_{i=1}^{n} \left(U_{X_i} \frac{\partial r}{\partial X_i} \right)^2 \right]^{1/2}$$

$N = B.S$ and therefore, $\frac{\partial N}{\partial B} = S = 50$ and $\frac{\partial N}{\partial S} = B = 10$.
Then

$$U_p = \left[U_B^2 \left(\frac{\partial N}{\partial B} \right)^2 + U_S^2 \left(\frac{\partial N}{\partial S} \right)^2 \right]^{1/2} = \left[(1 \times 50)^2 + (2 \times 10)^2 \right]^{1/2} = 53.8 = \sim 54 \text{ books}$$

$\therefore \ N = 500 \pm 54 \text{ books}.$

Example 1.4

In a tension test, the Young's modulus of a material is being measured. The maximum load was measured as 200 ± 2 N and the length of the gage section measured by a ruler is 50 ± 0.05 mm. The change in the gage-section length was 5 ± 0.02 mm. The width of the specimen is measured by a micrometer as 10 ± 0.03 mm. The thickness is measured by a micrometer as 2 ± 0.01 mm. What is the error (uncertainty) in the value of Young's modulus measured?

Solution:

Error in Young's modulus $E (= \sigma/\varepsilon)$ is:

$$U_E = \left[U_\sigma^2 \left(\frac{\partial E}{\partial \sigma} \right)^2 + U_\varepsilon^2 \left(\frac{\partial E}{\partial \varepsilon} \right)^2 \right]^{\frac{1}{2}}$$

$$= \left[U_\sigma^2 \left(\frac{1}{\varepsilon} \right)^2 + U_\varepsilon^2 \left(\frac{-\sigma}{\varepsilon^2} \right)^2 \right]^{1/2}$$

The uncertainty in stress and strain need to be determined.

Given:
$F = 2{,}000$ N and $U_F = 2$ N

$L = 50$ mm and $U_L = 0.05$ mm (half of resolution in the ruler)
$b = 10$ mm and $U_b = 0.03$ mm (half of resolution in the micrometer)
$t = 2$ mm and $U_t = 0.01$ mm (half of resolution in the micrometer)
$\Delta L = 5$ mm and $U_{\Delta L} = 0.02$ mm

$$\text{stress} \qquad \sigma = \frac{F}{A} = \frac{200}{20} = 10 \frac{N}{mm^2} = 10 \text{ MPa}$$

$$\text{Strain} \qquad \varepsilon = \frac{\Delta L}{L} = \frac{5}{50} = 0.1$$

$$\text{Young's modulus} \qquad E = \frac{\sigma}{\varepsilon} = \frac{10}{0.1} = 100 \text{ MPa}$$

Uncertainty in stress $\left(\sigma = \frac{F}{A}\right)$

$$U_\sigma = \left[U_F^2 \left(\frac{\partial \sigma}{\partial F}\right)^2 + U_A^2 \left(\frac{\partial \sigma}{\partial A}\right)^2 \right]^{\frac{1}{2}} = \left[U_F^2 \left(\frac{1}{A}\right)^2 + U_A^2 \left(\frac{-F}{A^2}\right)^2 \right]^{\frac{1}{2}}$$

$$= \left[\left(\frac{1}{20} \times 2\right)^2 + \left(-\frac{2000}{400^2} \times 0.116\right)^2 \right]^{\frac{1}{2}}$$

$$U_\sigma = 0.1$$

and uncertainty in area $(A = bt)$

$$U_A = \left[U_b^2 \left(\frac{\partial A}{\partial b}\right)^2 + U_t^2 \left(\frac{\partial A}{\partial t}\right)^2 \right]^{\frac{1}{2}} = \left[(tU_b)^2 + (bU_t)^2 \right]^{\frac{1}{2}}$$

$$= \left[(2 \times 0.03)^2 + (10 \times 0.01)^2 \right]^{\frac{1}{2}}$$

$$U_A = 0.116$$

Uncertainty in strain $\left(\varepsilon = \frac{\Delta L}{L}\right)$

$$U_\varepsilon = \left[U_{\Delta L}^2 \left(\frac{\partial \varepsilon}{\partial (\Delta L)}\right)^2 + U_L^2 \left(\frac{\partial \varepsilon}{\partial L}\right)^2 \right]^{\frac{1}{2}} = \left[U_{\Delta L}^2 \left(\frac{1}{L}\right)^2 + U_L^2 \left(\frac{-\Delta L}{L^2}\right)^2 \right]^{\frac{1}{2}}$$

$$= \left[\left(\frac{1}{50} \times 0.02\right)^2 + \left(\frac{5}{50^2} \times 0.05\right)^2 \right]^{\frac{1}{2}}$$

$$U_\varepsilon = 0.00041$$

Now we can determine uncertainty in Young's modulus from equations at the beginning of the solution as

$$U_E = \left[\left(\frac{1}{0.1} \times 0.1 \right)^2 + \left(\frac{10}{0.01} \times 0.0004 \right)^2 \right]^{\frac{1}{2}} = \pm 1.07 \text{ MPa}$$

The Young's modulus $E = 100 \pm 1.07$ MPa.

Example 1.5
A student team designed a solar electric vehicle. The dashboard contains a speedometer and a clock. When the car was traveling at a constant speed, the speedometer reads 70 ± 1 mph and the time measured over a distance for repeated trials was 2 ± 0.01 hr. What is the error in the distance traveled?

Solution:
Distance $= D$, Time $= t$ and Speed $= V$
Measured variable $(X_1 = t, X_2 = V)$
Results $=$ Distance travelled $(r = D)$
Uncertainties in the measured variables are:

$$U_{X_1} = U_t = \pm 0.01 \text{ hrs and } U_{X_2} = U_V = \pm 1 \text{ mph}$$

Uncertainty in the results U_r

$$U_r = \left[\sum_{i=1}^{n} \left(U_{X_i} \frac{\partial r}{\partial X_i} \right)^2 \right]^{\frac{1}{2}}$$

$$D = V \times t$$

the sensitivity of the distance traveled to time is $\frac{\partial D}{\partial t} = V = 70$ and
the sensitivity of the distance traveled to velocity is $\frac{\partial D}{\partial V} = t = 2$.
The uncertainty in the distance traveled is:

$$U_D = \left[U_V^2 \left(\frac{\partial D}{\partial V} \right)^2 + U_t^2 \left(\frac{\partial D}{\partial t} \right)^2 \right]^{\frac{1}{2}} = \left[1^2 \times 2^2 + (0.01)^2 (70)^2 \right]^{\frac{1}{2}}$$

$$= [4 + 0.49]^{\frac{1}{2}} = \sqrt{4.49} = 2.11 \text{ miles}$$

Therefore, the distance traveled (D) is $70 \times 2 \pm 2.12$

$$D = 140 \pm 2.12 \text{ miles}$$

Example 1.6

The elastic strain energy (the area under elastic portion of the stress–strain curve; see Fig. 1.7) for a certain material is given by $\delta = \frac{1}{2}\sigma\varepsilon$. If the load cell reads 100 ± 1 MPa and modulus measured is 200 ± 3 GPa, calculate the uncertainty in the strain energy.

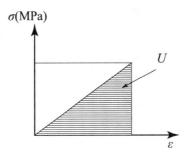

Figure 1.7: Elastic strain energy.

Solution:

$$\left. \begin{array}{l} \delta = \dfrac{1}{2}\sigma\varepsilon = \dfrac{1}{2}\dfrac{\sigma^2}{E} \\[2mm] \varepsilon = \dfrac{\sigma}{E} \end{array} \right\} \delta = \dfrac{1}{2}\dfrac{\sigma^2}{E}$$

$$\begin{aligned} U_\delta^2 &= \left[U_\sigma^2 \left(\frac{\partial\delta}{\partial\sigma} \right)^2 + U_E^2 \left(\frac{\partial\delta}{\partial E} \right)^2 \right] \\[2mm] &= \left[(1)^2 \left(\frac{\sigma}{E} \right)^2 + (3 \times 10^3)^2 \left(-\frac{\sigma^2}{2}\frac{1}{E^2} \right)^2 \right] \\[2mm] &= \left[(0.001)^2 \left(\frac{0.1}{200} \right)^2 + (3 \times 10^3)^2 \left(-\frac{\sigma^2}{2}\frac{1}{E^2} \right)^2 \right] \\[2mm] &= \left[(0.001)^2 \left(\frac{0.1}{200} \right)^2 + (3)^2 \left(-\frac{0.1^2}{2}\frac{1}{200^2} \right)^2 \right] \\[2mm] &= 0.000000625 \text{ GPa} = 0.000625 \text{ MPa} \end{aligned}$$

$$\delta = \frac{1}{2}\frac{100^2}{200 \times 1000} = \frac{1}{2} \times \frac{1}{20} = 0.025 \text{ MPa}$$

Uncertainty $= 0.025$ MPa ± 0.000625 MPa.

PART B: EXPERIMENT

1.5 DYNAMIC DATA ACQUISITION

1.5.1 OBJECTIVE

Dynamically acquire voltage signals from a Data Acquisition (DAQ) device using the Lab-VIEW development software and perform statistical analysis of the acquired signals.

There are four parts to this lab.

Part 1: Measurement of a fixed reference voltage using the DAQ and LabVIEW.

Part 2: Quantification and analysis of the accuracy associated with data acquisition using the DAQ.

Part 3: Estimation of strain in an object using a strain gage bonded to a feeler gage or a thin flexible metal strip.

    A. Use the strain gage in the Wheatstone-bridge configuration to allow strain measurement.

    B. Measure the diameter of an approximately circular object using strain measurement.

Part 4: Quantify uncertainty associated with estimating the strain and diameter of the object in Part 3.

1.5.2 BACKGROUND NEEDED FOR CONDUCTING THE LAB

Familiarity with LabVIEW programming is essential for all the labs. Useful LabVIEW instructional videos can be found on National Instruments website. A brief tutorial on how to construct a VI for a lab is also provided in the appendix in Section 1.10.

1.5.3 PRELAB QUESTIONS

Generate a VI that meets the requirements for Part 1 of Lab 1. Submit a pdf of a screen shot of the back panel of the VI. An appendix is provided that develops the VI needed (see Section 1.10).

1.5.4 EQUIPMENT AND RESOURCES NEEDED

List of suppliers for these items is provided at the end of this lab write up (Fig. 1.8).

- Laptop computer with LabVIEW installed

- Multi-function DAQ with USB cable

- Wire jumper kit

- Three 120 Ω resistors

- Breadboard

- AA battery and holder

- Strain gage mounted on a feeler gage or a thin metallic strip

- Measurement tools in lab (rulers, micrometers, etc.).

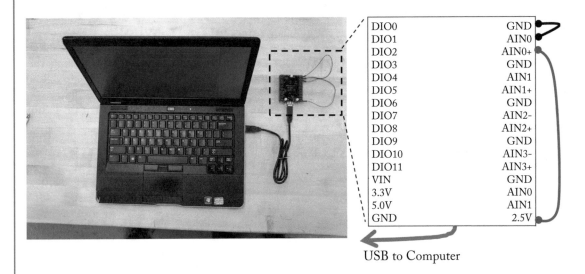

Figure 1.8: Tools and supplies required to conduct lab exercise.

1.6   PART 1: MEASUREMENT OF A FIXED REFERENCE VOLTAGE USING THE DAQ AND LABVIEW

1.6.1   PROBLEM STATEMENT

To examine the performance of the DAQ as it measures a fixed reference voltage in digital form with different gain windows (voltage ranges).

1.6.2   WHY ARE WE DOING THIS?

To understand and learn how voltage signals can be measured (and visualized) in the form of digital data using the DAQ and LabVIEW. Such voltage measurement (and the associated scatter in the data) can be used to measure a physical quantity (and its variability). In Part 3 of this lab, we will use such voltage measurements to estimate strain in an object (physical quantity) and then use mechanics equations to measure diameter (physical quantity) of an object and the uncertainty in the measured diameter.

1.6.3 REQUIRED LABVIEW PROGRAM (VI)

1. Write a LabVIEW program that uses the "Gain.vi" as a sub-vi that:

    (a) Calculates the signal mean and signal standard deviation, and indicates the results on the front panel of the acquired signal on AIN0 (in mV) as a function of time (in seconds). Recall that the DAQ is programed to capture data at the sample rate over the acquisition time (set by the user). The averages of this set of data are returned on the channel averages, right side of the SubVI. The raw data is retuned in the raw data array, top right of the SubVI.

    (b) Creates a dynamically updating X-Y graph scatter plot on the program front panel that graphs the statistical standard deviation of the acquired signal on AIN0 (in mV) as a function of time (in seconds).

    (c) Exports the data (voltage mean and standard deviation) to an Excel spreadsheet. Use this data to include a plot in the report.

2. When the program is functioning properly, perform an experiment where the channel voltage mean and standard deviation are monitored as the device sample window is changed over its range (via a front panel gain control) and record the data to a spreadsheet.

    Note: Take only meaningful data (include all sample windows that are meaningful, avoid saturation, if you leave any out, justify).

1.6.4 CONNECTIONS REQUIRED

See Fig. 1.9.

1. Connect a short jumper wire from the 2.5 V output terminal on the lower right side of the DAQ to the AIN0+ (**A**nalog **IN**put channel **0 positive**) terminal near the top right of the DAQ.

2. Connect a short jumper between any ground (GND) terminal and the AIN0- (**A**nalog **IN**put channel **0 negative**) terminal.

3. With these connections, the acquired (measured) voltage signal will be the difference between the positive and negative terminals, i.e., ((AIN0+)–(AIN0–)). The ground (GND) terminal represents the reference voltage (usually zero volts). Since the negative terminal (AIN0–) is connected to device ground and positive terminal (AIN0+) is connected to 2.5 V, the voltage signal measured is ~2.5 − 0.0 = 2.5 V. This is called a differential voltage measurement and will be the way all voltage signals will be measured in this laboratory. Remember, noise is associated with ALL voltage measurements, so you will NOT get an exact reading of 2.5 V, but all digital data collected will be close to it but scattered (random). Also remember we are using Channel 0 of the DAQ for acquisition of the data.

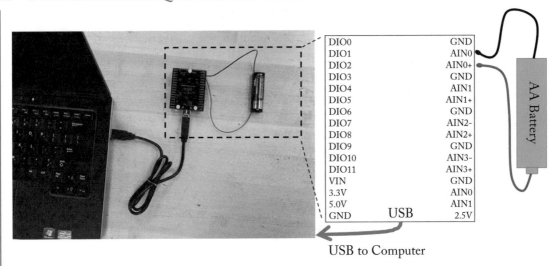

Figure 1.9: Illustration of wire connections to the computer from the back side of DAQ.

1.6.5   EXPERIMENTAL TASK FOR PART 1

1. Once the DAQ connections and LabVIEW VI are functional, follow the steps below to complete the Part 1 of this lab.

2. Power the DAQ by connecting it to the laptop with the USB cable provided. The green light on DAQ will flash indicating that it is powered. Wait for a few seconds until the laptop detects the DAQ.

3. On the front panel of the VI, set the sampling window on Channel 0 to ± 10 V. We are measuring a reference voltage of 2.5 V on channel 0 of the DAQ (see connections in Fig. 1.9).

4. Change the index of the index array block on VI to 1.

5. Read this step completely before you proceed. As the VI runs, monitor the channel voltage mean and standard deviation for 10 s. After 10 s, change the sampling window via the front panel to ±5 volts and continue to monitor the voltage mean and standard deviation. Repeat this procedure after every 10 s for four sample windows (±10 V, ±5 V, ±2.5 V, ±1.25 V). Once all sampling windows are recorded stop the VI.

6. Once the VI stops, the writetospreadsheettool.vi dialogue box will open. Save the data to a suitable location as Lab_x_Part_1.xls.

7. Disconnect the DAQ from laptop.

1.6.6   ISSUES TO BE DISCUSSED IN THE LAB REPORT FOR PART 1

1. Develop and present in the report two plots from the saved data: a voltage mean vs. time and a standard deviation of the voltage mean vs. time.

2. The reference voltage we were trying to measure on Channel 0 is approximately 2.5 volts with some noise. Can you see this from the voltage mean vs. time graph for all sampling windows? Comment on your observation.

3. In the second graph, observe the variation of standard deviation as you change the sampling windows at the intervals of 10 s. What happens? Why? If you observe zero standard deviation, explain what it means.

1.7   PART 2: QUANTIFICATION OF ACCURACY IN MEASUREMENTS MADE BY THE DAQ

1.7.1   PROBLEM STATEMENT

Quantify and analyze the accuracy associated with DAQ voltage measurements.

1.7.2   WHY ARE WE DOING THIS?

When we perform measurements in the lab, there are always some uncertainties associated with data collected. Bias (or constant errors) can be directly addressed with differential measurements and taring operations. Random noise in the measurements cannot be easily removed. Therefore, it is a standard practice to monitor, quantify, and report these random uncertainties during experimentation. In Part 2 we will develop a method to quantify this random noise associated with voltage measurements. It is expected that this method will be utilized when any voltage is being measured in Lab using the DAQ and reported in your lab reports.

1.7.3   REQUIRED LABVIEW PROGRAM

Open a blank VI.

1. Add a Gain SubVI (note: no "while-loop" will be used in this VI).

2. Add controls for acquisition time, Channel 0 sample window, and sample rate.

3. Add an index array to the data array and index channel 0 out.

4. Use a histogram tool and plot a histogram of the raw data.

5. Calculate the average and standard deviation of the raw data and report on the front panel.

6. Add a write to spreadsheet SubVI and save the channel 0 raw data.

7. Save the part 2 VI.

1.7.4   CONNECTIONS REQUIRED

As shown in Fig. 1.9, connect a battery (+ and − terminal) to the differential channel AIN0.

1.7.5   EXPERIMENTAL TASK FOR PART 2

Once the connections are made, follow the steps given below to finish Part 2.

1. Power the DAQ by connecting it to the laptop with the USB cable provided. The green light on DAQ will flash indicating that it is powered. Wait for few seconds until the laptop detects the DAQ.

2. Change the index of the index array block to 1 (if it is not).

3. Set the acquisition time to 3 s and leave the sample rate at the default 1,000 Hz.

4. Set the sampling window on Channel 0 to ±10 V.

5. Run the VI.

6. Once the VI stops, the writetospreadsheettool.vi dialogue box will open. Save the data to a suitable location as Lab_1_Part_2_1.xls.

7. Observe the histogram displayed; what is the distribution of the data?

8. Repeat steps 4, 5, 6, and 7 for smaller sample windows until saturation is detected.

9. Disconnect the DAQ from laptop.

1.7.6   ISSUES TO BE DISCUSSED IN THE LAB REPORT FOR PART 2

1. For each sampling window, develop the mean and standard deviation for 3 s of reading from the DAQ. Present these results in a table in the report. This is a quantification of the measurement error of the DAQ for the voltage measured with the DAQ set on the specific channel/gain window. Bias errors are not easily detected with this approach (and will be addressed with tarring operations) but the standard deviation is a good representation of the random noise in the measurement, provided the noise is Gaussian.

2. Plot a histogram for the ±10 V sample window data.

3. Make observations about the variation of the readings in each sample set. Is the variation random, or are some readings repeated? Why?

1.8   PART 3: ESTIMATION OF STRAIN IN AN OBJECT USING A STRAIN GAGE

1.8.1   PROBLEM STATEMENT

Use the stain gage (installed on a feeler gage) along with known resistors in the Wheatstone-bridge configuration to make strain measurement. Estimate the diameter of an object using this strain measurement.

1.8.2   WHY ARE WE DOING THIS?

Whenever an object with a strain gage installed on it is subjected to strain (applied loads), it causes a change in resistance of the gage. Such change in resistance is proportional to the applied strain and therefore can be used to estimate the magnitude of strain (this is how transducers work in general). Detailed discussion on "Theory of Strain Gages" will be provided in Lab 2. The Wheatstone-bridge configuration provides an effective way to measure and transform such small resistance changes into proportional voltage signal. Such voltage signals can be measured and analyzed using the DAQ along with LabVIEW. In Part 3, we use strain gage in a Wheatstone-bridge configuration and the DAQ to measure the strain in a thin metallic strip. The strain measurement is then used to estimate the diameter of an object around which the metal strip is wound.

1.8.3   BACKGROUND

A strain gage is used to measure strain in the feeler gage when bent around a circular object. The following equation is used to estimate strain (ε) based on the measurement of two voltages (output voltage V_G and supply voltage V_S) and the gage factor (G_f) of the strain gage

$$\varepsilon = \frac{4(V_G)}{V_s G_f} \tag{1.9}$$

The diameter of a large circular object can then be calculated using the equation

$$\varepsilon = \frac{t/2}{\rho} \tag{1.10}$$

Where t is the thickness of the feeler gage and ρ is the radius of curvature of the object.

1.8.4   REQUIRED LABVIEW VI

1. Use Gain.vi SubVI in a while loop similar to Part 1. We will measure V_G and V_S using the DAQ on Channels 0 and 1, respectively. Design the VI to do so. Include front panel indicators to display V_S and V_G.

2. TARE operation: Resistors used in this part are $R_1 = R_2 = R_3 = 120 \ \Omega$. The initial resistance of the strain gage is $R = 120 \ \Omega$. This means when the gage is NOT strained, the

Wheatstone-bridge is balanced, and ideally $V_G = 0$. But, in reality, the resistance of the strain gage may not be exactly 120 Ω and the connecting lead wires generally possess small but finite resistance. This causes an imbalance of the Wheatstone-bridge and gives rise to a non-zero value of V_G in step 1. This is a typical case of bias or systematic error. To avoid this, a TARE operation is used which is to manually subtract this non-zero value of V_G from that obtained in step 1 (ensure the feeler gage is not undergoing any strain). Display this tared V_G on front panel using indicator.

3. Using V_s and tared V_G acquired in steps 1 and 2 and knowing $R_1 = R_2 = R_3 = 120$ Ω, and the gage factor of the strain gage, program the VI to calculate the change in strain the strain gage is measuring. You may need to use simple numerical tools available in the LabVIEW.

4. Add the functionality to calculate the standard deviation of the measured voltages (V_G, V_s).

5. Include the ability to save tared V_G, V_s, the standard deviation of V_G, V_s, and strain as a function of time. You will need shift registers, initialize and build arrays, current time tool, and writetospreadsheet tools for this (similar to Part 1's VI).

6. Add graphs of quantities tared V_G, R_U, and strain V_s time for better visualization.

7. (Optional) Implement an expression to estimate the radius of curvature.

1.8.5   CONNECTIONS REQUIRED

A Wheatstone-bridge configuration will be implemented on a breadboard using three known resistors (R_1, R_2, and R_3 each with resistance 120 $\Omega \pm 0.1\%$) and a strain gage (with initial resistance of 120 Ω) bonded on a feeler gage acting as the unknown resistance (when bent). This arrangement is shown in Fig. 1.10. Place the gage in the R_U location of the bridge. Power the bridge with +5 volts from your DAQ.

To construct a Wheatstone-bridge: Consider each node as a row on the breadboard. (Note: each row on the breadboard has the same potential and they are all electrically connected.) Plug relevant connects at the node (numbered nodes shown in Fig. 1.10) into the row. Note that the unknown resistance is the strain gage on the feeler gage when the metal strip is bent.

Install jumpers for 5 V power (Wire to +5 volts on the DAQ, ground on the DAQ), as shown in Fig. 1.11.

Install jumpers for data acquisition with DAQ.

1.8.6   EXPERIMENTAL TASK FOR PART 3

The strain gage on the feeler gage (thin metallic strip) is now connected to Wheatstone-bridge, DAQ, and the computer, as shown in Fig. 1.13.

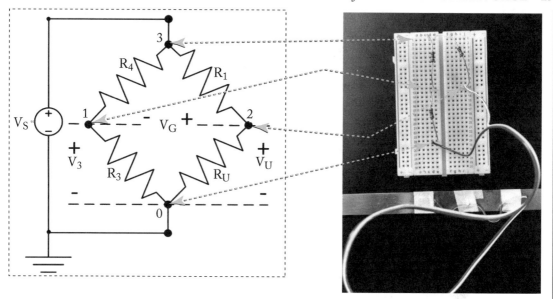

Figure 1.10: Schematic of the Wheatstone-bridge and equivalent breadboard connections with the resistors and strain gage.

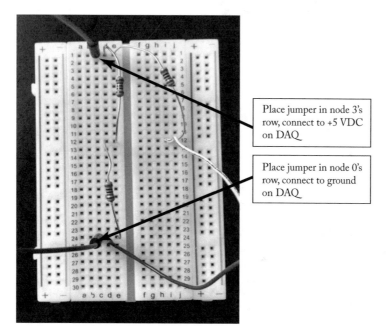

Place jumper in node 3's row, connect to +5 VDC on DAQ

Place jumper in node 0's row, connect to ground on DAQ

Figure 1.11: Installation of jumper cables to connect the Wheatstone-bridge to power supply and ground terminals (refer to Fig. 1.10 for node identification).

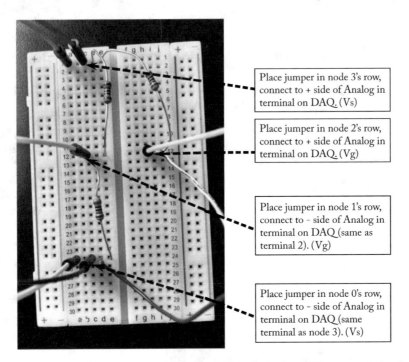

Place jumper in node 3's row, connect to + side of Analog in terminal on DAQ. (Vs)

Place jumper in node 2's row, connect to + side of Analog in terminal on DAQ. (Vg)

Place jumper in node 1's row, connect to - side of Analog in terminal on DAQ (same as terminal 2). (Vg)

Place jumper in node 0's row, connect to - side of Analog in terminal on DAQ (same as node 3). (Vs)

Figure 1.12: Connections for jumper cables from breadboard to Wheatstone-bridge and DAQ (refer to Fig. 1.10 for node identification).

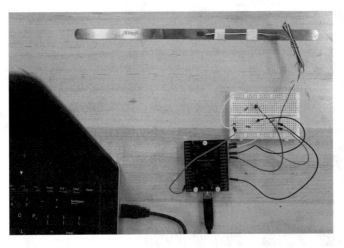

Figure 1.13: Wire connections from the strain gage on the metallic strip to the computer.

After the Wheatstone-bridge circuit and LabVIEW VI are verified, perform the following tasks and include the results and related discussions in your lab report. Report which sampling windows were used for measurement of V_s and V_G and justify.

1. Verify that VI yields the correct value for V_G with no strain applied (close to 120 Ω). You will perform a tare operation to accomplish this.

2. Manually strain the feeler gage "*in compression*" for ~10–15 s and record the range of variation seen in resistance R_U.

3. Measure the *diameter* of a circular object provided in the lab. Use a method that reduces errors.

4. Bend the feeler gage around the circumference of the object (see Fig. 1.14) while the Part 3 VI is running. Measure and record the strain induced and use it to calculate the diameter of the object. Ensure that the strain gage is not strained beyond its limits. You will need to use the relationship from bending of beams learned in MoM or use Eq. (1.10).

Figure 1.14: Illustration of feeler gage with strain gage bent around the circumference of a circular object.

1.8.7   ISSUES TO BE DISCUSSED IN THE LAB REPORT FOR PART 3

1. Include a plot of the variation of V_G vs. Time as the feeler gage is flexed in the report. Comment on the results. Does the voltage increase or decrease? Why?

2. Discuss about the tare operation you used in VI to get accurate value of V_G in the report.

3. Include a plot of the variation of V_G vs. Time as the feeler gage is flexed in the report. Comment on the results. Does the voltage increase or decrease? Why?

4. Report the method used and the resultant diameter found from Part 3 of the experimental task.

5. Report the diameter found from the use of the strain measurement.

6. Report the uncertainty found for each diameter, and discuss.

7. Discuss in the report what will happen if diameter of the object chosen is too small (~5 cm).

1.9 PART 4: UNCERTAINTY CALCULATIONS

1.9.1 PROBLEM STATEMENT

Develop the uncertainty in the two estimates of diameter.

Use the methodology from Part 2 (Section 1.7) to develop an estimate in the uncertainty in V_G and V_s. Propagate this uncertainty into estimates of resistance, strain and diameter accordingly. Develop and estimate the uncertainty of the *direct measure* of the diameter using a liner scale. In your report discuss the uncertainties estimated, using them to evaluate both schemes used to estimate diameter of the object measured.

1.9.2 ISSUES TO BE DISCUSSED IN THE LAB REPORT FOR PART 3

Include an appendix in the report that:

1. Details all values and uncertainties used in the uncertainty calculations (a table is appropriate).

2. Shows the application of the Root-Sum-Square approach (see Section 1.4) to propagating the uncertainty in strain. At this stage, it is not necessary to fully develop each partial derivative, but each partial taken should be symbolically represented.

1.9.3 EQUIPMENT REQUIREMENTS AND SOURCING

- Multi-function DAQ with USB cable: Minimum 2 channel 14 bit ADC, USB interface, LabVIEW support. Typical vendor: Out of The Box SADI DAQ,

  https://ootbrobotics.com/

- Wire jumper kit: 7″ male/male, 28 gage

  Typical vendors: Digikey, All Electronics

- Three 120 Ω resistors: Precision 1/4 watt resistors, ±.01%

  Typical vendors: Digikey

- Breadboard: 400 tie

  Typical part: Bud Industries BB-32621, typical vendor: Digikey

- AA battery (any supplier) and holder

  Typical vendors: Digikey, All Electronics

- Strain gage mounted on a feeler gage or a thin metallic strip:

  120 Ω uniaxial strain gage, typical vendor: Micro Measurement

  0.30 mm thick feeler gage, typical vendor: Mcmaster Carr

- Measurement tools in lab (rulers, micrometers, micrometers, etc.).

  Various suppliers

1.10   APPENDIX A: PART 1 – PREPARING VI

In this appendix, a brief presentation on how to construct a VI using LabVIEW programing is provided. For more in-depth details, the student is referred to LabVIEW program manual and instructions.

Lab 1 has three VI's required. The first VI must:

- Calculate the signal standard deviation and signal mean and indicate the results on the front panel of the acquired signal on AI0 (in mV).

- Create a dynamically updating X-Y graph scatter plot on the program front panel that graphs the statistical standard deviation of the acquired signal on AI0 (in mV) as a function of time (in seconds).

- Export the data (voltage, mean, standard deviation) to a .csv file (Excel, Matlab).

How?

1. Open LabVIEW, select "Create Project."

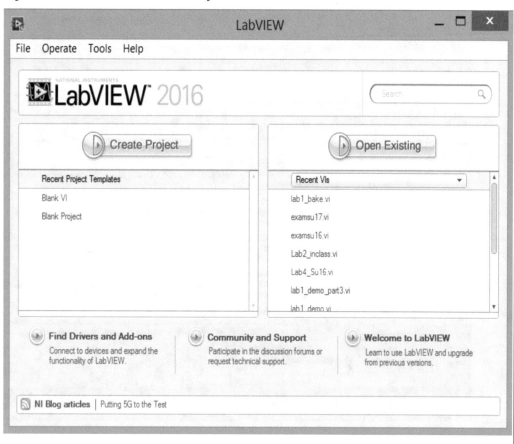

2. Select "Blank VI."

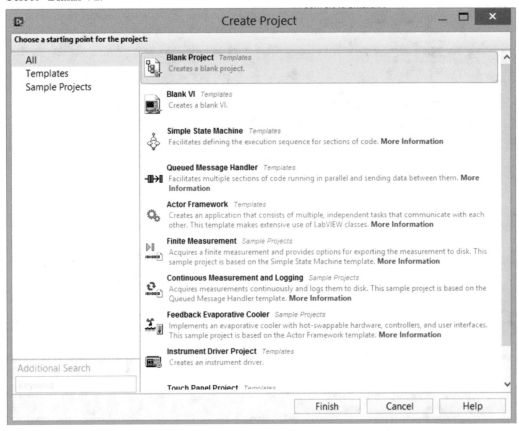

3. Front panel and block diagram.

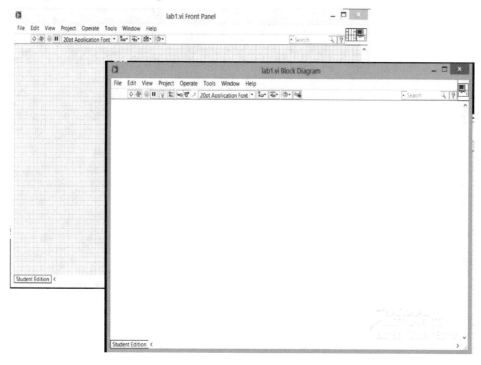

4. Start with the block diagram.

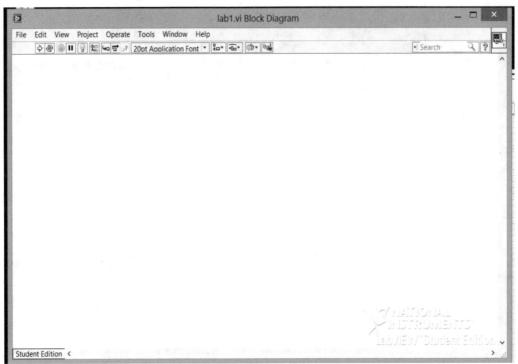

5. We want to sample over time, so we need to repeatedly call the data acquisition device.

    (a) There are several structures, use the While loop.

    (b) The structures pallet can be brought up with the right mouse button, and a click on the structures icon.

    (c) Select the While Loop, draw in block diagram.

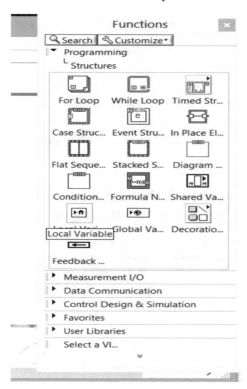

6. While loop: drag rectangle.

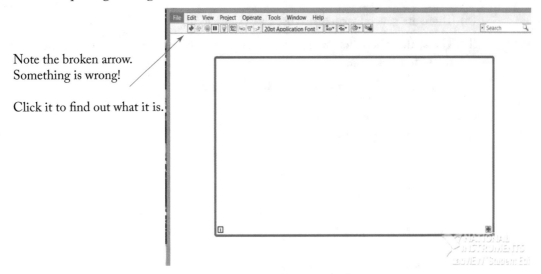

Note the broken arrow. Something is wrong!

Click it to find out what it is.

7. Add a control to stop the while loop when the user is finished with the data acquisition.

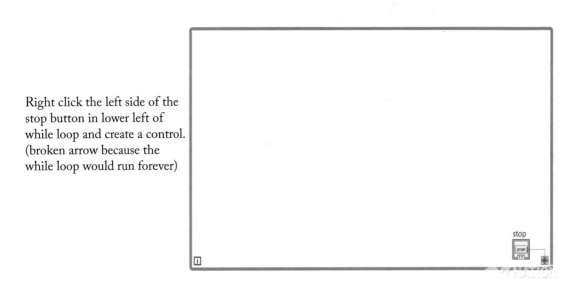

Right click the left side of the stop button in lower left of while loop and create a control. (broken arrow because the while loop would run forever)

8. Insert a SubVI to call DAQ.

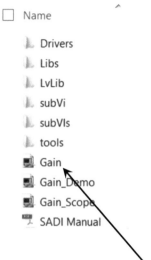

Right click, select VI, choose the "Gain" VI

9. Use the "Control h" key combination to see the interactive help on the Gain SubVI.

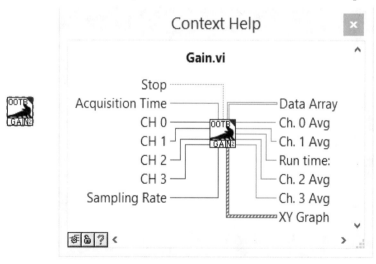

LabVIEW executes from left to right. Convention is that inputs are on the left of a SubVI, and outputs are on the right.

10. Add a wire between the while stop control and the Gain SubVI Stop control to shut down the DAQ when exiting the while loop.

11. Add indicators and controls to Gain.

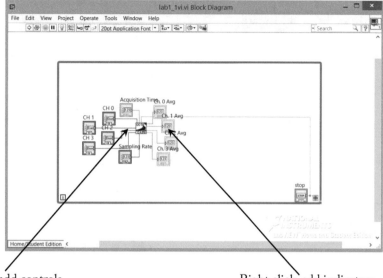

Right click, add controls   Right click, add indicators

Recall the first requirement: *Calculates the signal standard deviation and signal mean and indicates the results on the front panel of the acquired signal on AI0 (in mV) as a function of time (in seconds).*

(a) We need to calculate the standard deviation. How?

   i. → LabVIEW has many built in functions

(b) The raw data the daq collects is available in a $5 \times N$ array at the top. N is the product of the sample rate and acquisition time.

(c) We need the standard deviation of one channel or a subarray of the raw data.

12. Index Array: You can Quick Drop: keystroke control-space, search for "index array."

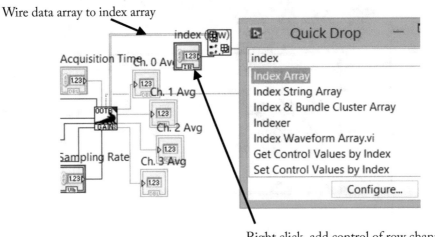

13. Find standard deviation SubVI.

    (a) You can Quick Drop: keystroke control-space, search for "std", and find standard deviation SubVI.

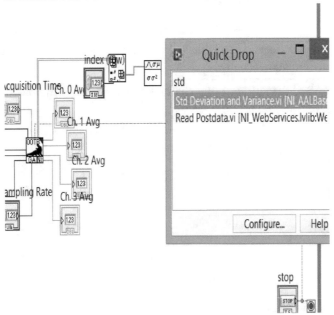

(b) Or you can right click, select search, upper right, then search std.

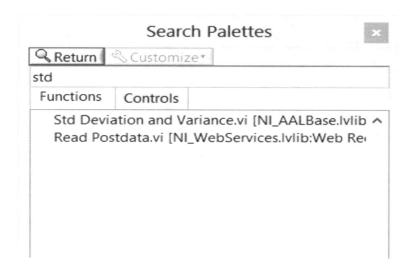

14. Select the channel going to the standard deviation tool.

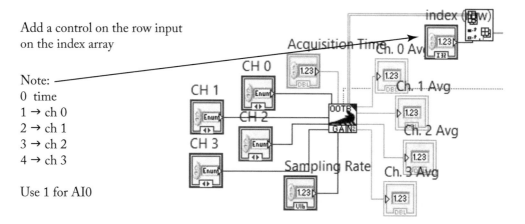

15. The output of the DAQ is in Volts. We want the standard deviation of the reading in mV.

(a) Tool pallet, numeric, multiply.

(b) Add a constant of 1,000 to multiply by.

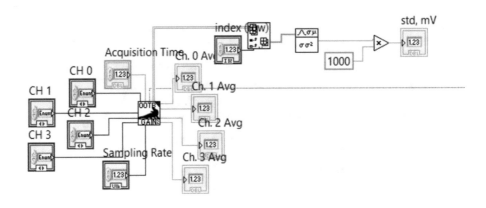

16. The front panel has the mean and standard deviation of channel 0 displayed.

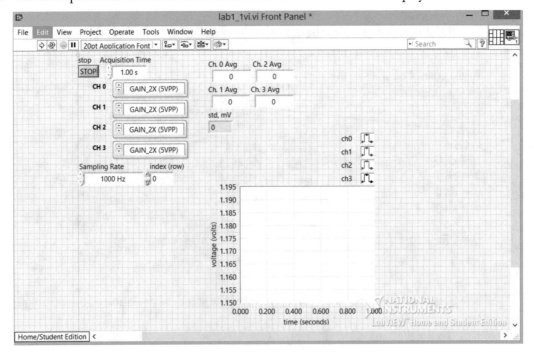

Part 1 works, but only shows the mean and standard deviation for the current iteration.

Recall the second requirement: *Create a dynamically updating X-Y graph scatter plot on the program front panel that graphs the statistical standard deviation of the acquired signal on AIO (in mV) as a function of time (in seconds).*

(a) Storage of data iteration to iteration.

(b) Shift registers and build arrays.

(c) Time.

17. Shift register.

(a) Note the newly created shift register is black—it has no memory associated with it, and will use unallocated memory if it is used as is. This is bad.

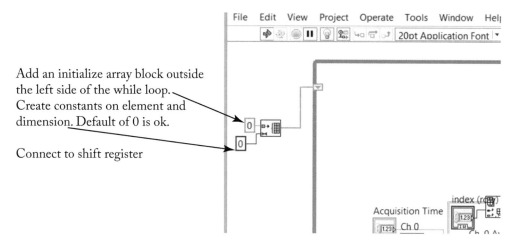

Add an initialize array block outside the left side of the while loop. Create constants on element and dimension. Default of 0 is ok.

Connect to shift register

18. Shift register, build array.

(a) Find a build array block (array pallet).

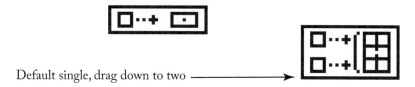

Default single, drag down to two

19. Shift register, build array, hook up standard deviation.

    (a) Add label for Standard Deviation Array Wire.

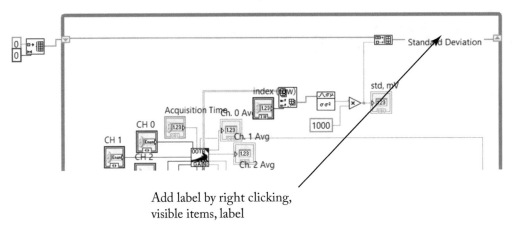

Add label by right clicking,
visible items, label

20. Shift register, build array.

    (a) Add mean. You can use the same initialize array.

        i. We need this for Part 3.

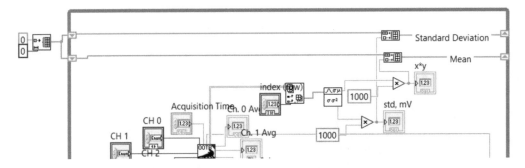

21. We need to plot mean vs. time. Establish elapsed time.

   (a) Class library has a "currentTimeTool" in the tools subdirectory.

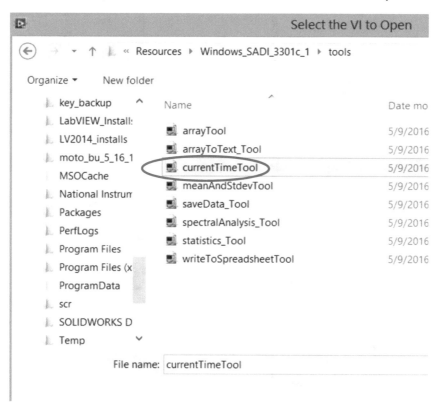

22. Place a "currentTimeTool" outside of the while loop. This will read the time before the loop starts running, giving the initial time.

Drag a wire to the left side of the while loop. The orange square is called a tunnel.

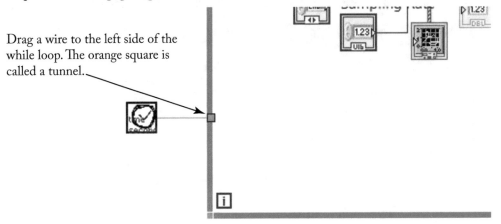

23. Place a "currentTimeTool" inside of the while loop. This will read the time at the start of each loop.

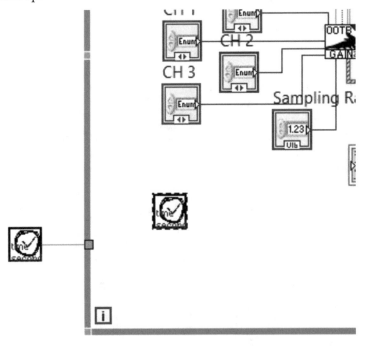

24. Take the initial time away from the current time to get the elapsed time.

   (a) Use numeric pallet for subtraction.

   (b) Add a shift register and build an array of the elapsed time values.

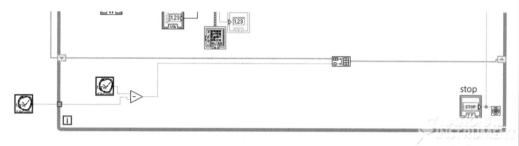

25. A scatter plot is needed for the second requirement.

   (a) Use the one configured as an indicator on the bottom of the Gain SubVI.

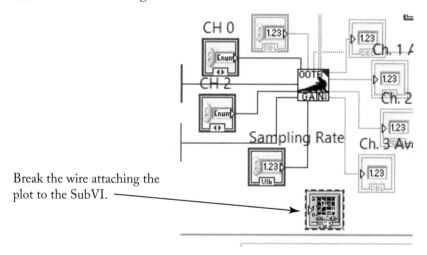

Break the wire attaching the plot to the SubVI.

26. Create a Bundle of the time as the independent variable and the standard deviation of the voltage from channel 0 as the dependent variable.

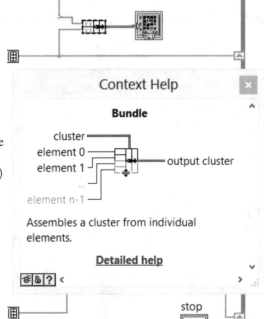

A bundle is used to associate the independent variable (top) with the dependent variable (bottom)

Recall the third requirement: *Exports the data (voltage, mean, standard deviation) to an Excel spreadsheet.*

- We need to assemble several sets of data.
    - Build array outside the while loop, right side.
    - Assemble the three single-dimensional arrays build in the while loop in to one $3 \times m$ array.
        * Note: use a new build array, or turn off concatenation.
    - The build arrays used to assemble scalars into 1D arrays concatenate the scalar to the end of the array. This is implicitly set by LabVIEW when you wire the build array.
- We need to write them to a file.
    - Use the writeToSpreadsheetTool provided in the class library.

27. Lab 1 Part 1 VI: Block diagram.

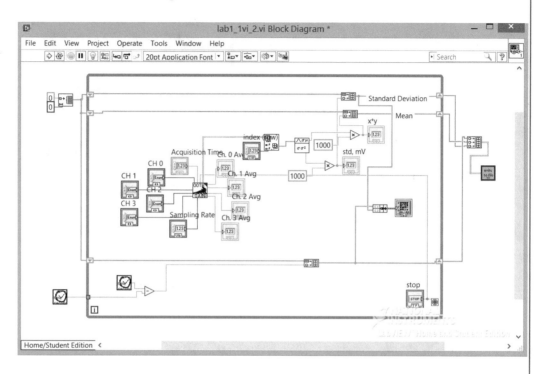

Lab 1 Part 2 VI: Open a blank VI and

- Add a Gain SubVI (note: no while loop will be used in this VI).

- Add controls for acquisition time, Channel 0 sample window, and sample rate.

- Add a constant on the stop function and set to true.

- Add an index array to the data array and index channel 0 out.

- Use a histogram tool and plot a histogram of the raw data.

- Calculate the average, and standard deviation of the raw data and report on the front panel.

- Add a write-to-spreadsheet SubVI and save the channel 0 raw data.

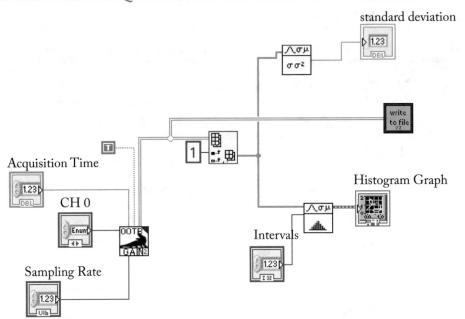

1.11 APPENDIX B: LAB REPORT FORMAT

A critical skill an engineer should possess is the ability to document their technical work. These lab exercises are designed to lead to a lab report that details the student's efforts, results, and conclusions. Accurately and concisely conveying this information is as important as generating it.

The lab reports submitted are to follow the IEEE transactions format:

`https://ieeeauthor.wpengine.com/wp-content/uploads/Transactions-template-and-instructions-on-how-to-create-your-article.doc`

A brief description of basic elements follows.

Typical sections:

	Abstract
	Index Terms
I.	Introduction
II.	Procedure
III.	Results
IV.	Discussion
V.	Conclusion
	Appendices
	References

1.11.1 ABSTRACT

- Abstract is a one- (maybe two-) paragraph summary of the report.

- An abstract should have three components:

  (1) a clear statement of the objectives.

  (2) background information required to understand how the objectives were achieved; and

  (3) a brief summary (quantitative) of the important results. Items (2) and (3) should support the objectives.

- The abstract exists independently from the rest of the paper.

- It should not refer to anything in the paper.

- It is suggested to do it last.

1.11.2 INDEX TERMS

- These are terms that could be used in a search that would return your paper.

- Three or four are sufficient.

- Must be in alphabetical order.

1.11.3 INTRODUCTION

- The introduction has two main components.

  (1) A clear statement of the objectives.
  (2) Background information required to understand how the objectives were achieved.

- A brief summary of the important results:

  The introduction exists independently from the abstract and should be written as if the abstract does not exist. Although the introduction has similar components as the abstract, it should have greater detail.

- The introduction should not be a tutorial on the topic.

- It should include information the reader needs to understand the details of what you did to achieve the objective.

- Generally, no more than three paragraphs are required for a good introduction.

  – A paragraph introducing the topic and the challenge being addressed.
  – A paragraph establishing the background theory or previous work.
  – A paragraph stating the specific objectives/hypotheses/aims of the present study.

1.11.4 PROCEDURE

- The procedure section, often called the Materials and Methods section, is a sufficiently detailed explanation of what you did that a motivated reader could reproduce your results.

  – Be sure to report what you did in a consistent voice (third person) and tense (usually past tense).
  – You may cite specification sheets, class notes, and assignment documents for minute details, but be sure the reader can understand all the important parts of your activity from your written description.

- Note any changes to the published write-up.

- Note any deviations for the published write-up in the experiment.

- Do not copy directly from the instructions, paraphrase, condense, and reference.

1.11.5 RESULTS

- The results section MUST include a written description of the key observations and findings of the exercise.

- You must provide the reader with a written high-level tour of your results that allows them to look at and make sense of your graphical and tabular information.

- Tables and graphs should be referred to in parentheses in support of a statement, rather than as the subject of a sentence.

  - For example, the sentence "Increasing loop times significantly increased the amount of step-response overshoot (Fig. 1)" is more active and definitive than the sentence "Figure 1 shows the relationship between loop times and step-response overshoot."

- Generally, there should be at least one sentence (or paragraph) describing the key findings for each of the study aims/objectives.

- The Results section should contain no interpretation of the results or comparison with previous work—that belongs in the Discussion section.

1.11.6 DISCUSSION

- The Discussion section should contain at least three main parts.

  - The first paragraph should provide a concise restatement of the study context, objectives, and key results.

- Subsequent paragraphs should interpret study findings (what do they mean?) and relate them to previously published work or established engineering principles.

  - The lab instructions include discussion questions. Answer them here.

    * Make it easy for the reader to find the answers to these questions.
    * Answer all of the questions.

  - Results go in the Results section, all other calculations go in the Discussion section.

    * These could be graphs and tables, or individual values.

- Finally, at least one paragraph should highlight the key technical, experimental, or procedural limitations of the study and establish the scope over which the study findings might apply (e.g., the results apply only to similar systems or to every engineering system in the universe).

- This limitation paragraph can be the last of the Discussion section, or it can be the second paragraph of the Discussion section, depending upon your style preference.

1.11.7 CONCLUSION

- The Conclusion section generally will be one concise paragraph.

    - Summarize the key findings of the present work.
    - State the importance or impact of the findings on the field of application.
    - Suggesting next steps or future efforts to reach higher levels of performance or impact.
    - Support the conclusion by references results and/or discussion items.
    - Use your uncertainty analysis to support your conclusions.
    - Include observations about improvements in the process to improve the results.

1.11.8 REFERENCES

- Include references for any cited work.

    - Do not put an item in this section if you do not cite it in the report.

- The lab instructions were most likely used in the generation of the report.

1.11.9 APPENDICES

- It is suggested that an appendix be used to present the uncertainty analysis utilized in the report.

1.11.10 GENERAL FORMAT

Figures

- Referred to in text with, even if at the beginning of a sentence, Fig. 1.

- Should appear after mentioned in text, and every figure used should be mentioned in text.

- Caption starts with: Fig. 1. Caption:

    - Two spaces between Fig. 1 and Caption.
    - Used to explain figure significance (why does this figure matter?).

- Do not put box around figure.

- DO NOT CONNECT DISCRETE DATA POINTS

    - This implies a functional relationship.
    - A trend line or function is allowable.

- Label axes and give units.

Tables

- Table should not appear before it is mentioned in text.

- Referred to in text with: Table I.

- Numbered with Roman numerals.

- Title goes at top.

- Caption/notes (if needed) go at bottom.

- Units must be included.

- Table format is to be consistent throughout paper.

Citations

- Citation appear in square brackets (ex. [1]).

- Cite figures in the figure label or in the first figure reference.

- When citing two sources at once the following notation is used [1], [2] if citing three or more sources [1]–[#].

- Capitalize only the first word in the paper title, except for proper nouns and element symbols.

- Give all the author's names; do **not** use "et al."

Equations

- Number is right justified, equation centered in the column.

    - Use a table (one row, three columns).
    - Place the equation in the center column, center justified.
    - Place the equation number in the right column, right justified.
    - Turn off the borders of the table.

- Cite with (1); only spell out Equation at beginning of a sentence.

- Make sure equation is legible.

Significant figures

- *Do not over report.*

- Use the uncertainty in a measured value to establish the significant figures reported for that measurement.

- If the uncertainty in a result does not support more than X significant figures, do not put more than X significant figures in.

General

- Include units for all reported values.

- Use scientific notation where appropriate; use exponents, not "^".

- Keep units consistent, and do not mix SI with English.

- Define all abbreviations on first use.

- Delete all template text.

- PROOFREAD THE WORK.

LABORATORY 2

Design and Build a Transducer to Measure the Weight of an Object

PART A: THEORY

2.1 CANTILEVER BEAM, STRAIN GAGES, AND WHEATSTONE-BRIDGE

In this laboratory the student will design and build a transducer (or sensor) to measure the weight of a beverage can (or water bottle) using a cantilever beam with a strain gage bonded on its top surface. We utilize cantilever beam theory to build this sensor. A thin metal strip with a bonded strain gage is used for this purpose. There are two components to the theory: (i) mechanical part that deals with flexural loading of a cantilever beam and (ii) electrical part that deals with resistance strain gages in a Wheatstone-bridge (W-bridge).

2.2 CANTILEVER BEAM THEORY

A cantilever beam of length L_o, height h, and width b is bonded with a strain gage on the top surface at a distance L from the weight W which acts at the free-end, as shown in Fig. 2.1. The average stress and average strain along the cross-section (AA') at the center of the gage can be derived from beam theory.

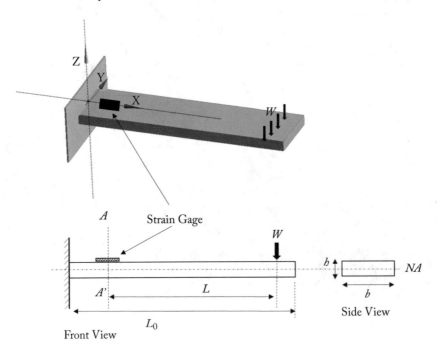

Figure 2.1: Cantilever beam with a strain gage bonded on its surface and a concentrated load W acting at the free-end.

The flexural stress in a cantilever beam, along x-axis, at section AA' is given by

$$\sigma_x = \frac{MY}{I} \tag{2.1}$$

where, I is the moment of inertia and the bending moment M about the y-axis is

$$M = W.L \qquad \text{N-mm} \tag{2.2}$$

$$I = \frac{bh^3}{12} \, \text{mm}^4 \qquad \text{and} \qquad c = \frac{h}{2} \, \text{mm} \tag{2.3}$$

Therefore, stress on the top surface (at $Y = h/2$) at section AA' is

$$\sigma_x = \frac{MY}{I} = \frac{(WL)\left(\frac{h}{2}\right)}{\left(\frac{bh^3}{12}\right)} = \frac{6WL}{bh^2} \qquad \frac{\text{N}}{\text{mm}^2} \qquad \text{(MPa)} \tag{2.4}$$

Thus,

$$\sigma_x = \frac{6WL}{bh^2} \quad \text{MPa} \tag{2.5}$$

Note that the stress is along the length of the beam (x-direction), although the load is acting vertical to the beam. For simplicity, we assume that this is the only stress acting in the beam and hence use 1D Hooke's law to determine strain at any point in the beam. Thus, strain at AA' is

$$\sigma_x = E\varepsilon_x \Rightarrow \varepsilon_x = \frac{\sigma_x}{E} \tag{2.6}$$

where, E is elastic modulus of the beam material.

Substituting Eq. (2.5) into Eq. (2.6), the strain is given by

$$\varepsilon_x = \varepsilon = \frac{6WL}{Ebh^2} \tag{2.7}$$

For simplicity we will use $\varepsilon_x = \varepsilon$, with the understanding that the strain is along the axial direction. Since the dimensions of the beam (b and h) and the Young's modulus (E) of the beam are fixed, the strain is purely a function of weight (W) and the location of the gage. Since the strain gage is bonded at a fixed distance L on a given beam with known dimensions and material properties, the strain is purely a function of weight of the bottle.

Rearranging Eq. (2.7), we get

$$W = \varepsilon_x \left(E \frac{bh^2}{6L} \right) \quad \text{or} \quad W = \sigma_x \left(\frac{bh^2}{6L} \right) \tag{2.8}$$

Thus, W is a function of stress (or strain) only.

Using the above theory, let us first establish the upper limit for the sensor measurement. As a designer you need to establish this limit so that you are aware of the range over which the sensor is capable of making the desired measurements accurately. If you exceed this limit, either the sensor will get damaged (giving faulty measurements from then on) or the structure on which the sensor rests will permanently get damaged (exceeds yield stress of material) making it incapable of acting as a sensor. Thus, there are two considerations for establishing the maximum safe weight the sensor can measure: (i) related to the maximum limit (load or stress) the beam material can withstand and (ii) related to the maximum limit the strain gage can measure (or stretch elastically). If any one of them is exceeded, the sensor becomes faulty. You can address this issue by answering the following questions (use Eq. (2.8) to answer the questions).

Questions:

1. Knowing the dimensions of the beam (e.g., $L = 200$ mm, $b = 25$ mm, $h = 3$ mm, and $E = 70$ GPa), determine the maximum weight (W_{max}) that can be supported by the beam if the yield strength (σ_{yield}) of the beam material is 240 MPa.

2. If a strain gage can measure no more than 5% strain, can it still measure the above W_{max} correctly? Or, what is the safe weight the beam can hold without exceeding the strain gage limit?

From the above answers, one can estimate the maximum weight the beam can measure without either yielding the bar or damaging the strain gage.

2.3   STRAIN GAGES AND WHEATSTONE-BRIDGE

2.3.1   STRAIN GAGE THEORY

A strain gage is an electrical conducting resistant wire wounded tightly in a specific pattern on a backing material. It works on the principle of change in resistance when stretched or compressed axially under mechanical loads or subjected to temperature changes. The resistance (R) of a wire is given by

$$R = \frac{\rho l}{A} \tag{2.9}$$

where ρ is specific resistant of the wire material (a material property), l is the length of the wire and A is the cross section of the wire.

When a mechanical load is applied (or temperature changes) a wire elongates and the change in resistance is given by

$$dR = \frac{l}{A}d\rho + \frac{\rho}{A}dl - \frac{\rho l}{A^2}dA \tag{2.10}$$

Dividing both sides by Eq. (2.9)

$$\frac{dR}{R} = \frac{d\rho}{\rho} + \frac{dl}{l} - \frac{dA}{A} \tag{2.11}$$

Note that the term $\frac{dl}{l}$ is axial strain (ε_x) and $\frac{dA}{A}$ is the lateral strain which can be calculated as

$$A = \frac{\pi}{4}D^2 \tag{2.12}$$

hence,

$$dA = \frac{\pi}{4}2D\,dD \qquad \boxed{\frac{dA}{A} = \frac{2dD}{D} = 2\varepsilon_y}$$

Recall that Poisson's ratio is

$$v = \frac{-\text{lateral strain}}{\text{axial strain}} = -\frac{\varepsilon_y}{\varepsilon_x} \qquad \rightarrow \qquad \varepsilon_y = -v\varepsilon_x \tag{2.13}$$

From Eqs. (2.11)–(2.13)

$$\frac{dA}{A} = 2\varepsilon_y = -2v\varepsilon_x = -2v\frac{dl}{l}$$

By rewriting Eq. (2.11)

$$\frac{dR}{R} = \frac{d\rho}{\rho} + \varepsilon_x + 2v\varepsilon_x = \frac{d\rho}{\rho} + \varepsilon_x(1 + 2v) \quad \text{where,} \quad -1 < v < 0.5 \; (v = 0.3) \tag{2.14}$$

Therefore,

$$\frac{dR/R}{\varepsilon_x} = \frac{d\rho/\rho}{\varepsilon_x} + 1 + 2v, \text{ if } v \text{ is } 0.5 \text{ (maximum value)} \rightarrow \frac{dR/R}{\varepsilon_x} = \frac{d\rho/\rho}{\varepsilon_x} + 2 \tag{2.15}$$

Here, $\frac{dR/R}{\varepsilon_x}$ is defined as sensitivity of the gage alloy, S_A.

It should be noted that for most materials, S_A is linear as shown in Fig. 2.2. If a high specific resistance alloy is chosen, then $\frac{d\rho}{\rho} \rightarrow 0$ in Eq. (2.11), which now reduces to

$$\frac{dR/R}{\varepsilon_x} \approx 2.0 \tag{2.16}$$

However, the strain gage wire may also contain sensitivity to lateral strain and shear strain. All these sensitivities are grouped into one term and called the *Gage factor* (G_f) and thus the Eq. (2.15) can be written as

$$\frac{dR/R}{\varepsilon_x} = G_f \qquad \text{or} \qquad \frac{dR}{R} = G_f\varepsilon_x \tag{2.17}$$

If R is the initial resistance of a strain gage under no strain ($\varepsilon = 0$) and R_U is the unknown (new) resistance under applied load ($\varepsilon \neq 0$) of a strain gage bonded to a structure, the above equation can be written as

$$R_U = R + \varepsilon G_f R \tag{2.18}$$

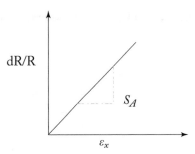

Figure 2.2: Sensitivity of the gage alloy.

The unknown resistance R_U needs to be determined or measured. From Eq. (2.17), when the gage is not strained ($\varepsilon = 0$), $R_u = R$ (the original resistance of the gage). The term (G_f) is the gage factor that represents the sensitivity of the gage to strain. The gages we typically use in this lab have $R = 120 \pm 0.3\% \, \Omega$ with a gage factor (G_f) of 2.1 $\pm$ 0.5%. The manufacturer supplies this value when strain gages are purchased.

A strain gage being simply an electrical conducting wire, it should be used in conjunction with a circuit to obtain a strain value. Here we use the W-bridge circuit and read voltage drop across a strain gage.

2.3.2   WHEATSTONE-BRIDGE

The W-bridge is constructed of two simple circuits wired in parallel, with the simple circuit named a voltage divider. As the name implies, the voltage divider circuit takes an input voltage and divides it among several circuit elements. Figure 2.3 shows an example of a voltage divider using resistors, R_1 and R_2.

In Fig. 2.3, the source voltage V_S represents the total electrical potential difference in the circuit. Kirchoff's Voltage Law (KVL) states that the sum of voltages around a closed loop must equal zero. If the nodal voltages are measured going clockwise around the voltage divider (e.g., node 2 → 1, node 1 → 0, and node 0 → 2), the resulting KVL equation is $V_1 + V_2 - V_s = 0$, i.e.,

$$V_s = V_1 + V_2 \tag{2.19}$$

Ohm's Law states that the voltage (V) across an element is proportional to the current (I) flowing through it. The proportionality constant (R) is the resistance of the element. This relationship is given by

$$V = IR \tag{2.20}$$

Kirchoff's Current Law (KCL) states that the sum of currents flowing in and out of a node (node 1 for example) must be zero. For the circuit in Fig. 2.3, the KCL is written as

$$I_1 - I_2 = 0 \therefore I_1 = I_2 = I \tag{2.21}$$

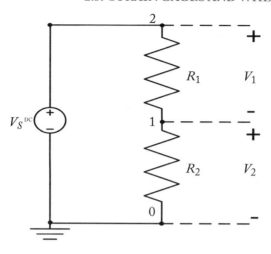

Figure 2.3: Example of a voltage divider circuit.

Stated alternatively, in a voltage divider, all elements see equal current. When the resistors are connected in series as in Fig. 2.3, the total resistance is the sum of the individual resistances, i.e.,

$$R = R_1 + R_2 \tag{2.22}$$

For this example, assume that the resistance of R_2 is unknown while the resistance of R_1 is known. Substituting Eqs. (2.19), (2.21), and (2.22) into Eq. (2.20), we obtain

$$V_s = I\,(R_1 + R_1) \tag{2.23}$$

In this assignment, the constraint is to only use voltage measurements to determine the unknown resistance value, so the current term (I) in Eq. (2.23) must be eliminated. Rearranging Eq. (2.23) to solve for the current we obtain,

$$I = V_s \left(\frac{1}{R_1 + R_2} \right) \tag{2.24}$$

Substituting the Ohm's Law equation of the unknown resistor, $V_2 = IR_2$, the current term in Eq. (2.24) is eliminated and the expression for a two-element voltage divider in terms of only voltages and resistances is given by,

$$V_2 = V_s \left(\frac{R_2}{R_1 + R_2} \right) \tag{2.25}$$

If the value of the unknown resistor $R_2 (= R_U)$ is desired, Eq. (2.25) can be rearranged as

$$R_2 = \frac{R_1}{\left(\frac{V_s}{V_2} - 1 \right)} \tag{2.26}$$

In (2.26), R_1 is the known resistor, V_s is the known supply voltage, and the measured voltage across the unknown resistor is V_2.

A refinement of the above method is to use two voltage dividers connected in parallel and make a differential voltage measurement between the two dividers, V_G, to calculate an unknown resistance, R_U, where all other resistances are known. The unknown resistor can be a strain gage with known initial resistance but unknown resistance when it is bonded to a structure and a load is applied to the structure. This differential divider circuit is called a W-bridge, and its configuration is shown in Fig. 2.4.

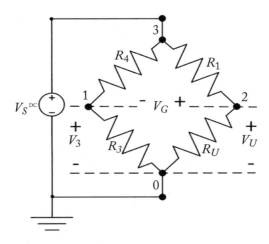

Figure 2.4: Two voltage dividers connected in parallel to form a W-bridge.

The voltages in each divider can be found by applying Eq. (2.25) to each half of Fig. 2.4, with the expression shown below.

$$V_3 = V_s \left(\frac{R_3}{R_3 + R_4} \right) \quad \text{and} \quad V_U = V_s \left(\frac{R_U}{R_1 + R_U} \right) \tag{2.27}$$

The known resistance divider circuit is from node 3 → node 1 → node 0. The unknown resistance divider circuit is from node 3 → node 2 → node 0.

The divider differential voltage, V_G, is simply the difference between each voltage and it is given by:

$$V_G = V_U - V_2 = V_s \left(\frac{R_U}{R_1 + R_U} \right) - V_s \left(\frac{R_3}{R_3 + R_4} \right) \tag{2.28}$$

$$V_G = V_s \left(\frac{R_U}{R_1 + R_U} - \frac{R_3}{R_3 + R_4} \right) \tag{2.29}$$

Solving (2.29) for the unknown resistance, R_U, yields

$$R_U = \frac{R_1 \left(\frac{V_G}{V_s} + \frac{R_3}{R_3+R_4} \right)}{\left(1 - \left(\frac{V_G}{V_S} + \frac{R_3}{R_3+R_4} \right) \right)} \tag{2.30}$$

Equation (2.30) gives a value of unknown resistance, given a known source voltage V_S, three known resistances R_1, R_3, and R_4, and a single voltage measurement between the two voltage dividers, V_G. In general, it is customary to choose all the known resistances to be the same, i.e., $R_1 = R_3 = R_4 = R = R_o$ initial resistance ($\varepsilon = 0$) value of the strain gage. This ensures that the product of resistances in opposite arms of the bridge is the same, i.e., $R_1 R_3 = R_2 R_4$. Under these conditions, the current across the nodes 1 and 2 is zero, a condition called "null balance." When one of the resistances is a strain gage and is experiencing strain, the voltage across nodes 1 and 2 is not zero any more. Under these conditions, the V_G (or ΔV) is given by

$$V_G / V_s = G_f \varepsilon / 4 \tag{2.31}$$

In many instances, it may be necessary to use more than one strain gage in the W-bridge. If strain gages are used in all the four arms of the bridge, then V_G is given by:

$$V_G / V_s = G_f / 4 \left(\varepsilon_1 - \varepsilon_2 + \varepsilon_3 - \varepsilon_4 \right) \tag{2.32}$$

Note alternating $+$ and $-$ signs for the strains in the above equation. By suitably arranging the strain gages in the circuit, one could double or quadruple the signal from the W-bridge. This issue will be further discussed in future laboratory experiments. Although the W-bridge seems overly complex, it is actually quite a useful way to inter-relate changes in resistance and voltage and will be used in the all the experiments where strain gages are used.

In summary, by measuring the change in differential voltage V_G and source voltage V_s, one can measure and monitor the changes in the unknown resistor R_U or the strain gage. In this part, we will use strain gage as an unknown resistance (R_U) placed in the branch between nodes 2 and 0. Whenever the gage is strained, its resistance (R_U) changes which can be measured using Eq. (2.32). In all the rest of the labs, we will use ΔV instead of V_G because it represents the difference in voltage measured when a load is applied compared to its unstressed condition.

Due to economic and time considerations, often a single-strain gage is used in many laboratory experiments. In such conditions, the above equation reduces to (assuming that the strain gage is located in the "arm 1" of the bridge),

$$\Delta V / V_s = G_f \varepsilon_1 / 4 \quad \text{or} \quad G_f \varepsilon / 4 \quad \rightarrow \quad \boxed{\varepsilon = \frac{4(\Delta V)}{V_S G_f}} \tag{2.33}$$

The above equation will be used in many of the experiments where a single strain gage is used.

The bonded strain gage on the cantilever beam is now connected to a W-bridge for strain measurement. Lead wires are attached to the gage and these wires are connected to one of the arms of the W-bridge. Since only one strain gage is used (in arm #2 of W-bridge), the remaining three arms of the W-bridge are fixed resistances. This configuration is called a "quarter bridge." The W-bridge with four fixed resistors is shown in Figs. 2.5a and b, the W-bridge with quarter bridge arrangement with R_2 as the strain gage resistance is shown. In Fig. 2.5b, the lead wires are shown with their own resistance R_L. The leadwire resistance can sometimes offset the null-balance condition; to be discussed later.

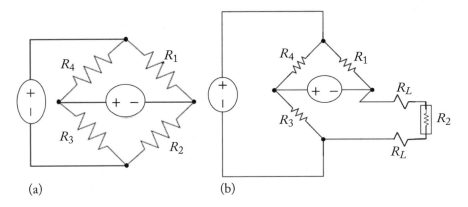

(a)   (b)

Figure 2.5: Schematics of a W-bridge strain gage circuit with (a) all fixed resistances and (b) quarter-bridge with two lead wires of resistance R_L.

Recall that the output voltage (ΔV) from a full W-bridge (when all the resistances are replaced by strain gages in the bridge) is given by

$$\frac{\Delta V}{V_s} = \frac{G_f}{4} (\varepsilon_1 - \varepsilon_2 + \varepsilon_3 - \varepsilon_4) \tag{2.34}$$

Because only one strain gage is connected in R_2 position and the rest are fixed resistances in Fig. 2.5a, there is no induced strain due to applied loads in these resistances, i.e., $\varepsilon_1 = \varepsilon_3 = \varepsilon_4 = 0$.

Therefore, the Eq. (2.34) for 1/4-bridge becomes

$$\frac{\Delta V}{V_s} = \frac{G_f}{4} (\varepsilon_2) \quad \text{or} \quad \frac{\Delta V}{V_s} = \frac{G_f}{4} (\varepsilon) \tag{2.35}$$

The negative sign is not considered for now. For generalization, we have used $\varepsilon_2 = \varepsilon$. One can also connect the strain gage in any other three arms of the W-bridge and the relationship remains the same.

Note that in Fig. 2.5b the strain gage is connected to the bridge with two long lead wires. Each of these electrical wires will have their own resistance indicated by R_L. The original null-balance condition (without lead wires) is given by:

$$R_1 R_3 = R_2 R_4 \tag{2.36}$$

In the presence of lead wires, the resistance in the strain gaged arm of the W-bridge is now changed to $R_2 + 2R_L$ and hence the null balance condition is not satisfied even when no load is applied on the cantilever beam, i.e.,

$$R_4(R_2 + 2R_L) \neq R_1 R_3 \tag{2.37}$$

Therefore, we resort to three-wire attachment of the strain gage as, shown in Fig. 2.6.

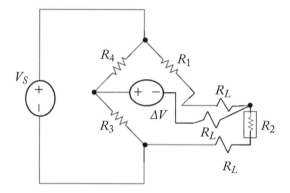

Figure 2.6: Three-wire "quarter-bridge" W-bridge circuit.

In this arrangement, we use another lead wire (R_L) and connect this to the central voltage-divider circuit where ΔV is measured. We also move the junction point to a new location as shown where R_2 meets with two lead wires. With this split in lead wire resistances, the null-balance condition is fully restored:

$$R_4(R_2 + R_L) = R_3(R_1 + R_L) \tag{2.38}$$

Thus, we can now start to measure strain induced by the strain gage upon application of load on the cantilever beam.

With this arrangement the output of the bridge is

$$\frac{\Delta V}{V_s} = \frac{G_f}{4}(\varepsilon_x) \tag{2.39}$$

Rearranging, the strain is given by

$$\varepsilon_x = \frac{4(\Delta V)}{V_s G_f} \tag{2.40}$$

Combining Eqs. (2.40) and (2.8), we can now calculate the weight of the bottle as:

$$W = \left(\frac{Ebh^2}{6L}\right)\varepsilon_x = \left(\frac{Ebh^2}{6L}\right)\frac{4(\Delta V)}{V_s G_f} \tag{2.41}$$

i.e.,

$$W = f(\Delta V) \tag{2.42}$$

Experimentally, the output of the bridge is (ΔV) (not strain) and can be measured using strain gages in a Wheatstone-bridge configuration and DAQ. Here,

ΔV—differential voltage from no-load to current load (weight)
V_s—source/excitation voltage
G_f—strain gage factor.

The output (ΔV) of the W-bridge is now connected to an amplifier ($\times 220 \pm 0.1\%$) and then run through one of the channels in the DAQ. The measurement is read on the Lab-VIEW VI on your monitor. This arrangement is shown in Fig. 2.7. Thus, the measured signal (ΔV_{amp}) is now an amplified (by 220 ×) voltage output of the W-bridge. Remember that each channel in the DAQ has a bias voltage (few mv) which you have measured in Lab 1 for Channel 0. This voltage is also included in the total output voltage ΔV_{amp}.

2.4   CALIBRATION OF THE TRANSDUCER

Once the theory of cantilever beam and strain gage circuits are understood, and the gage is bonded to the beam, a calibration is performed on the transducer. Recall from Eq. (2.41) and Fig. 2.7 that the output of the bridge is not strain but amplified ΔV (volts), i.e., ΔV_{amp}. So, we need to find out the relationship between the output $(\Delta V_{amp}) = (\Delta V) \times$ gain, and the input W (weight of the bottle). This procedure is called calibration and basically it provides how much ΔV_{amp} is generated for a unit weight. Also, notice from the Eq. (2.41) that all the variables, i.e., E, b, h, L, G_f, and V_s, are all known and so W is a function of ΔV only. It is a linear relationship. Now we want to ensure that the performance of the transducer is linear by conducting a calibration procedure.

In the laboratory, we use a commercial-off-the-shelf (COTS) scale to weigh some known weights. The weight of each block is indicated on them. The reason for re-measuring them using a scale is that the value indicated on each block is not exact and the measured weight deviates from the actual weight (variability in weights produced by the manufacturer). This range or variability must be quantified. The measurement process (scale) also has an uncertainty associated with it and we will assess how this error propagates into the final answer (i.e., weight of the bottle being measured).

The procedure for calibration is as follows.

1. Place a known weight at the free-end of the cantilever beam (your transducer) and measure the output voltage. Recall that the output voltage is passed through an amplifier (V_{amp}) and this amplified voltage is being measured. Repeat this process with larger weights, ensuring that the maximum weight you use is less than the $W_{\max}$ determined earlier by answering the questions in Section 2.2. The amplified signal you used is centered at ~2.5 V and so

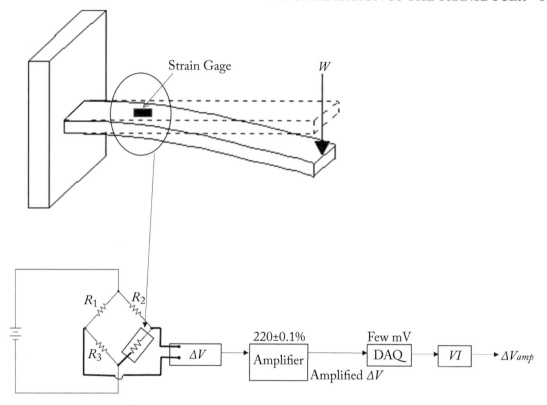

Figure 2.7: W-bridge connected to the strain gage on the cantilever beam and with differential amplifier gain.

for zero weight the signal should be around 2.5 V $\pm$ noise. Therefore, you may choose to tare the signal at the beginning or subtract 2.5 V from all the subsequent readings. The calibration should be performed with ΔV_{amp}, so the y-intercept is almost zero. Repeat these measurements for 5–7 weights. These values can now be plotted. The expected results (as per Eq. (2.41) is shown in Fig. 2.8).

The data in general falls on a straight line (see Eqs. (2.41) and (2.42)), as shown in Fig. 2.8, i.e., $Y = mx + C$. If the tared transducer signal is used, the C term will be zero by definition of the tarring operation. With this calibrated curve, we can determine the weight of any object placed on the beam at the same location where the calibration-weights were placed by measuring ΔV_{amp} from the bridge.

W	$\Delta Vamp$
0	0
50	0.02
.	.
.	.
.	.
450	0.1
500	0.15
550	0.2

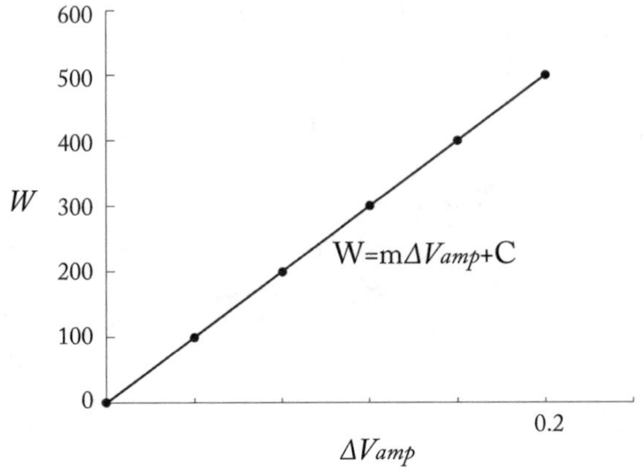

Figure 2.8: Calibration curve establishing the linear relationship between ΔV_{amp} and weight.

2.5 DETERMINE THE WEIGHT OF THE BOTTLE USING THE MOM METHOD

Using Eq.(2.41), measuring the dimensions of the beam, and assuming appropriate values for G_f and E, one should calculate the weight of the bottle 10 times (note V_s and ΔV_{amp} vary slightly for each measurement). Recall that because of noise in the instrument and measurement uncertainty, the values will be slightly different each time. Calculate average and standard deviation for the measurement. Report the values of weight of the full bottle.

Question:

1. Now you have calculated the weight of the bottle from the calibration curve method (Section 2.4) and MOM method (Section 2.5). Which one is a better measurement and why?

2.6 QUANTIFY UNCERTAINTY

2.6.1 CALIBRATION CURVE METHOD (CCM)

In this method, the linear relationship ($Y = mx + C$) established earlier in calibration method (Section 2.4) is used and uncertainty is quantified. From Fig. 2.8, we have

$$W = m\Delta V_{amp} + C \tag{2.43}$$

assuming $C = 0$, the sources of uncertainty in the measured weight are: Slope m and ΔV_{amp}.

The uncertainty in weight is given by

$$U_W = \sqrt{\left(\frac{\partial W}{\partial m}\right)^2 U_m^2 + \left(\frac{\partial W}{\partial \Delta V_{amp}}\right)^2 U_{\Delta V_{amp}}^2}$$

$$= \sqrt{\left(\Delta V_{amp}\right)^2 U_m^2 + (m)^2 U_{\Delta V_{amp}}^2} \qquad (2.44)$$

For calculation of maximum uncertainty, pick all maximum values. For full bottle, the ΔV_{amp} is maximum. The uncertainty in ΔV_{amp}, i.e., $U_{\Delta V_{amp}}$ is discussed in the following section (Uncertainty in ΔV_{amp}). Values for m and uncertainty in slope, i.e., $U_m = U_{slope}$ are obtained from the Monte Carlo (M-C) simulations (discussed in the section below).

Uncertainty in ΔV_{amp}

ΔV_{amp} is the tared reading from an amplifier representing the bridge output of a 1/4-bridge strain gage installed on a cantilever beam. The uncertainty in ΔV_{amp} can be estimated by sampling V_{amp} and developing the standard deviation of the readings. This uncertainty can then be propagated into an uncertainty of ΔV_{amp} using the RSS method. Recall that $\Delta V_{amp} = V_{amp} - \Delta V_{amp}$. There is uncertainty in both measurements caused by random error in the process.

Any uncertainty in the gain of the amplifier is addressed by the calibration method. This approach assumes that the voltage from the strain gage/amplifier pair is linear with respect to the force applied at the end of the beam. Thermal issues associated with the strain gage are neglected.

Uncertainty in Slope, U_m, from Monte Carlo Simulations

The uncertainty in the slope of Fig. 2.8 is determined from a least squares fit of several data points. Each data point represents the transducer's output in ΔV_{amp} for a given weight. Note that there is uncertainty in ΔV_{amp} as well as in the evaluation of the given weight. These uncertainties (spread in data) are indicated as lines with arrows in Fig. 2.9. A M-C simulation is utilized to estimate the propagation of these uncertainties into the uncertainty of the slope. (Background instructions on how to write a program to conduct M-C simulations is provided in the appendix in Section 2.10.)

Assume that the error in the COTS scale is 0.1 g, i.e., the range for each weight is

$$W_{\max} = W + 0.1\,\mathrm{g}$$
$$W_{\min} = W - 0.1\,\mathrm{g}$$

In calibration method, you used standard weights and measured output voltage from your transducer. Let us say you used five standard weights (~50 g, ~100 g, ~150 g, ...). Each scale-generated weight has an error of 0.1 g, i.e., $U_w = 0.1$ g.

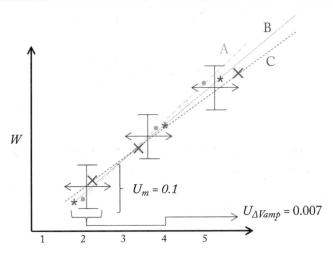

Figure 2.9: Illustration of uncertainty in slope m between W and ΔV_{amp} plot.

Assume that you have taken ΔV_{amp} for each weight at a certain sampling rate (e.g., 20 points/s). So, calculate average ΔV_{amp} and standard deviation, i.e., $U_{\Delta V_{amp}}$. Let us say, for illustration sake, the calculated $U_{\Delta V_{amp}}$ for each weight is around 7 mV or 0.007 V.

For each weight and for each ΔV_{amp}, there is an uncertainty (range) and this uncertainty can be represented as a box on W vs. ΔV_{amp} plot, as shown in Fig. 2.9. Any value in this box is a feasible value and there are infinite possibilities. Similar boxes are repeated at each data point. So one can write infinite lines going through these boxes. For example, three such possible lines are shown with slightly different slopes. M-C simulations pick a random number from each box for different set of weights in the W vs. ΔV_{amp} space and generate thousands of lines to calculate the average slope (m) and standard deviation (U_m) for the slope.

You are provided an excel program for M-C simulations. Your input to the program is five sets of weights ($Y_1, Y_2, \ldots Y_5$) you have used to calibrate the curve and their corresponding ΔV_{amp} values ($X_1, X_2, \ldots X_5$). These are input in the 1st row of the Excel sheet. In the 2nd row you will provide the uncertainty range for each of these values. It is quite possible that the uncertainty could be the same for all five values of weight or ΔV_{amp}. The M-C table supplied to you has been designed to generate a large set of values around the average and standard deviation of each set of points centered in the box. Once the values of five weights, corresponding ΔV_{amp} values and the uncertainty values for each are entered into the Excel sheet, it provides the average slope (m) of all these lines and the slope uncertainty of the line, U_m.

For each fixed weight (e.g., 50 g, 100 g ...) collect at least 100 readings of ΔV_{amp}. The sample rate is up to the student to decide: 20 pts/s, 100 pts/s, or 500 pts/s. A 95% confidence interval should be developed for the ΔV_{amp} signal to quantify the uncertainty in that signal.

Since the weights are fixed vales (e.g., 50 g, 100 g, ...) as measured on a weighing scale, the $Y_1, \ldots Y_5$ values are fixed. Also, the error on each weight is also fixed as indicated on the weighing scale (0.1 g) and so all the $U_{Y1}, \ldots U_{Y5}$ have the same values. Also note that each weight is not the value indicated on the weighing block, but should be measured on the weighing scale in the lab. For example, 50 g may read 49.7 g and 100 g may read 100.4 g. These exact values should be used as $Y_1 \ldots Y_5$.

After making the ΔV_{amp} measurements for each weight, use them for $X_1, \ldots X_5$ (i.e., $V_{amp\ 1}, V_{amp\ 2}, \ldots V_{amp\ 5}$) and the estimated uncertainty in the voltage for $U_{X1}, U_{X2}, \ldots U_{X5}$ (i.e., $U_{\Delta V_{amp\ 1}}, U_{\Delta V_{amp\ 2}}, U_{\Delta V_{amp\ 5}}$)in Table 2.1. Use the measured weights for Y_1 to Y_5, and the uncertainty in the scale for $U_{Y1} \ldots U_{Y5}$. The program generates 4,000 sets of points and gives average slope and uncertainty in slope (U_m).

Your starting Excel sheet will look like Table 2.1. You need to fill in the V_{amp} values (X-axis values and their uncertainties).

Table 2.1: Starting Excel sheet

X_1	X_2	X_3	X_4	X_5	Y_1	Y_2	Y_3	Y_4	Y_5
					49.7	100.4	159.6	199.2	250.5
U_{X1}	U_{X2}	U_{X3}	U_{X4}	U_{X5}	U_{Y1}	U_{Y2}	U_{Y3}	U_{Y4}	U_{Y5}
					0.1	0.1	0.1	0.1	0.1

Uncertainty in MOM Method

In the previous section we have discussed the uncertainty in the weight measured using the calibration method. One can also calculate the weight using the MOM. Recall from Eq. (2.8) that the weight placed at the end of the cantilever beam is given by

$$W = \sigma_x \left(\frac{bh^2}{6L} \right) \tag{2.8}$$

Knowing uncertainty in each parameter in the equation, we can calculate the uncertainty in weight which is given by

$$U_W = \sqrt{\left(\frac{\partial W}{\partial \sigma_x} \right)^2 U_{\sigma_x}^2 + \left(\frac{\partial W}{\partial b} \right)^2 U_b^2 + \left(\frac{\partial W}{\partial h} \right)^2 U_h^2 + \left(\frac{\partial W}{\partial L} \right)^2 U_L^2} \tag{2.45}$$

While the partial differentials are easy to solve, the uncertainties in dimensions $U_b, U_h,$ and U_L and stress (U_{σ_x}) need to be evaluated. For the uncertainties in dimensions it is common practice to take half the resolution of the instrument used for measurement of dimensions $b, h,$ and L. For example, if a ruler with 0.5 mm divisions is chosen to measure length of the beam, then the uncertainty in length (U_L) is 0.25 mm. But how do we get uncertainty in stress, U_{σ_x}?

Stress is given by $\sigma_x = E\varepsilon_x$. Now use the same equation to get uncertainty in stress. i.e.,

$$U_{\sigma_x} = \sqrt{\left(\frac{\partial \sigma_x}{\partial E}\right)^2 U_E^2 + \left(\frac{\partial \sigma_x}{\partial \varepsilon_x}\right)^2 U_{\varepsilon_x}^2} \tag{2.46}$$

Uncertainty in Young's modulus, U_E, is either obtained from manufacturer's website or from M-C simulations (not for this laboratory). For uncertainty in strain, U_{ε_x}, we use the equation $\varepsilon = \frac{4\Delta V}{V_s G_f}$ and calculate

$$U_{\varepsilon_x} = \sqrt{\left(\frac{\partial \varepsilon_x}{\partial \Delta V}\right)^2 (U_{\Delta V})^2 + \left(\frac{\partial \varepsilon_x}{\partial V_s}\right)^2 (U_{V_s})^2 + \left(\frac{\partial \varepsilon_x}{\partial G_f}\right)^2 U_{G_f}^2} \tag{2.47}$$

Here the uncertainty in gage factor U_{G_f} and uncertainty in source voltage U_{V_s} are supplied by the manufacturer of each. Uncertainty in voltage output $U_{\Delta V}$ is given in Eq. (2.20). By plugging in Eqs. (2.46) and (2.47) into Eq. (2.45), we can get the uncertainty in weight using the MOM method.

2.7   USE OF MULTIPLE-STRAIN GAGES ON THE CANTILEVER BEAM AND IN THE WHEATSTONE-BRIDGE

Thus far, we have discussed use of a single-strain gage on a cantilever beam in a quarter (1/4) bridge arrangement to measure strain. One should be aware of multiple strain gage use to enhance the signal to noise ratio from the sensor (i.e., cantilever beam). The positioning of strain gages in the W-bridge as well as on the mechanical structure (the cantilever beam) must be carefully considered to achieve higher signal. This can be achieved by careful consideration of the output of a W-bridge given by Eq. (2.34), i.e.,

$$\frac{\Delta V}{V_s} = \frac{G_f}{4}(\varepsilon_1 - \varepsilon_2 + \varepsilon_3 - \varepsilon_4) \tag{2.34}$$

2.7.1   HALF-BRIDGE (1/2-BRIDGE)

Four possibilities exist and we will illustrate two of them. The student should work out the other two cases (posed as questions at the end of each scenario).

Case 1: The intent here is to double the signal by taking advantage of alternating positive and negative signs in Eq. (2.34). Two strain gages are bonded on the opposite surfaces of the cantilever beam (top and bottom) at the same distance from the end and connected in adjacent arms of the W-bridge. Note that when the weight W is acting downward, as shown in Fig. 2.10, the gage on the top surface reads tensile ($+$) strain and the bottom one gives compressive ($-$)

strain, but the magnitudes are equal. If the strain gages are placed in arms 1 and 2 of the bridge, the output of the circuit is given by

$$\frac{\Delta V}{V_s} = \frac{G_f}{4}(\varepsilon_1 - (-\varepsilon_2) + 0 - 0) = \frac{2\varepsilon_1 G_f}{4} = \frac{G_f \varepsilon_x}{2} \rightarrow$$

twice the signals compared to the output of 1/4-bridge $\left(\frac{G_f \varepsilon_x}{4}\right)$.

Question:

1. What is the bridge output when the strain gages are arranged in opposite arms of the W-bridge and placed exactly, as shown in Fig. 2.10?

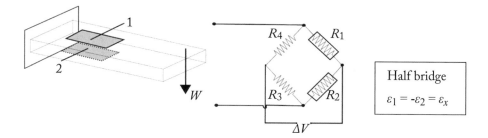

Figure 2.10: Schematic of the W-bridge arrangement when two strain gages are bonded on opposite surfaces of the beam (half bridge) and arranged in adjacent arms of the bridge.

Case 2: If we choose to place the two strain gages on the same side of the beam (top surface) next to each other and place them in opposite arms of the bridge, as shown in Fig. 2.11, both give the same positive (tensile) strain (ε_x). So, we place them on the opposite arms of the W-bridge.

The output now is given by

$$\frac{\Delta V}{V_s} = \frac{G_f}{4}(\varepsilon_1 - 0 + \varepsilon_3 - 0) = \frac{G_f(2\varepsilon_x)}{4} = \frac{G_f \varepsilon_x}{2}$$

Note that in both Cases (1) and (2), we get twice the ΔV for the same V_s and W.

Question:

1. What is the bridge output when the strain gages are arranged in adjacent arms of the W-bridge?

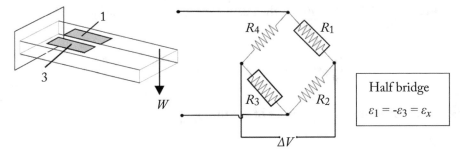

Figure 2.11: Schematic of the cantilever beam with two adjacent gages on the same surface placed in opposite arms of the W-bridge (half bridge).

2.7.2   FULL-BRIDGE

Here we use four strain gages—two on the top surface and two on the bottom surface of the beam exactly below the above two strain gages and arrange them appropriately to quadruple the output from the bridge. Students should try this exercise.

2.8   MICROMETER

While measuring small dimensions (such as thickness of a beam), micrometer is often used. It is an ingenious device for measurement of length at fine resolution. It consists of three scales: sleeve, thimble, and Vernier, as shown in Fig. 2.12.

Each scale is further divided into finer divisions. The maximum length that can be measured is given on the sleeve and is also indicated on the handle. For example, 0–1″. Similarly, the sensitivity of the instrument is also given next to it as 0.0001″. In the following, how this sensitivity is calculated is illustrated.

Sensitivity = Max. possible measurement/number of divisions on that scale.

For the micrometer shown here, the sensitivity of each scale is

1. Sleeve scale → 1.0/40 = 0.025 in.

2. Thimble scale → 0.025/25 = 0.001 in.

3. Vernier scale → 0.001/10 = 0.0001 in.

One complete revolution of thimble advances the spindle through 0.025 in on sleeve scale. Each thimble has 25 divisions and hence the sensitivity of the thimble is 0.025/25 = 0.001 in. There are also ten horizontal lines on the sleeve. So, the sensitivity of the Vernier scale is 0.001/10 = 0.0001 in while measuring the dimensions of an object, the line marker on the Vernier that

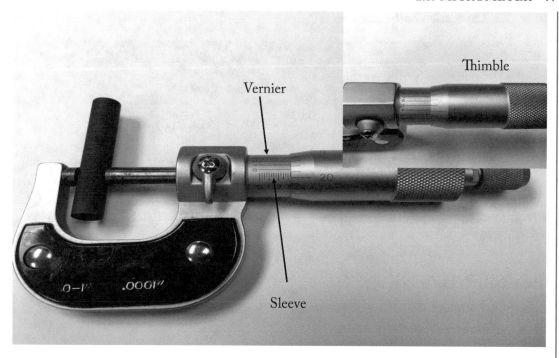

Figure 2.12: Micrometer with its sleeve scale, thimble scale, and Vernier scale.

coincides with the thimble scale, measurement should be considered. Practice measuring different objects using a micrometer to get familiarized with these basic concepts. Let us consider an example of measuring the outside diameter of a circular object such as the one shown in Fig. 2.12. Assume that the reading on each of the scales is given, as shown in Table 2.2. The relevant calculations are shown and the final calculation of the diameter is shown in Table 2.2.

Table 2.2: Calculation of the diameter

Scale	#Increment	Sensitivity	Reading (in)
Reading on the sleeve	3	$\frac{0.1}{4} = 0.0250$	$(3 \times 0.1) + (1 \times 0.0250) = 0.3250$
Thimble	21	$\frac{0.025}{25} = 0.001$	$21 \times 0.001 = 0.021$
Vernier line coincides exactly with a thimble line at division	5	$\frac{0.001}{10} = 0.0001$	$5 \times 0.0001 = 0.0005$
Total Reading			$0.3250 + 0.021 + 0.0005 = 0.3465$

REFERENCES

[1] J.W. Dally and W.F. Riley, *Experimental Stress Analysis*, McGraw Hill.

PART B: EXPERIMENT

2.9   CANTILEVER BEAM, STRAIN MEASUREMENT, AND UNCERTAINTY

2.9.1   OBJECTIVE

This lab is designed to demonstrate the use of a strain gage as a transducer to measure various weights placed at the free-end of a cantilevered aluminum beam. This lab is to be completed in two weeks. In Week 1, the student will bond a strain gage and in Week 2 the instrumented beam is held in a cantilever configuration, calibrated, and then used to weigh unknown weights. Bond the strain gage following the instructions provided by Vishay Strain Gage Installation Manual (Instruction Bulletin B-27-14, `http://www.vishaypg.com`) and any other pertinent information.

2.9.2   PRELAB PREPARATION

1. Develop a parametric representation of the stress at an arbitrary location along a cantilever beam for an arbitrary load.

    (a) This could be in LabVIEW, Excel, Matlab, by hand, etc.

    (b) You will use this in your LabVIEW program and in lab.

2. Recall the relationship between stress and strain. The material used in the bar is specified to be 6061 T6 Aluminum, with a yield point of approximately 240 MPa and a tensile modulus of elasticity of approximately 69 GPa.

2.9.3   EQUIPMENT AND SUPPLIES NEEDED

See Fig. 2.13.

• Laptop computer with LabVIEW installed.

• A multi-function DAQ.

• Power supply.

• Instrumented cantilever beam (assembled by the student) and support structure.

• Strain gage amplifier/power supply with W-bridge completion circuit.

• Calipers, micrometer.

• Various calibration weights.

• Commercial weighing scale to measure weights.

• Can of soda or water.

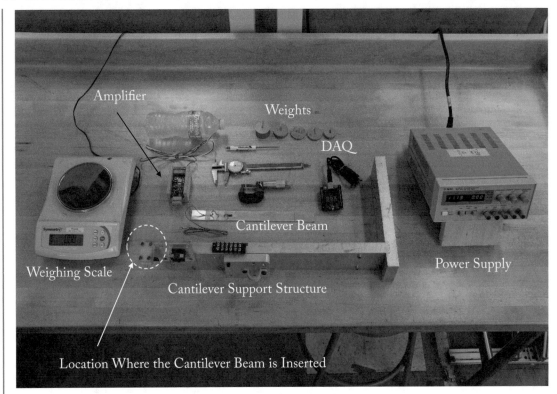

Figure 2.13: Equipment and supplies needed for conducting the experiment.

2.9.4   PROBLEM STATEMENT

Construct an instrumented cantilever beam to be used as a scale (transducer) and utilize it to determine the weight of a can of soda. Determine the weight of your average gulp, and the standard deviation of your average gulp. Perform these measurements in two ways: (1) by using the calibration curve for your beam based on known weights and (2) by using the gage factor for the strain gage, directly assessing the beam strain and stress, and then computing the applied load (use MOM and the strain gage circuit theory).

2.9.5   REQUIRED LABVIEW PROGRAM (VI)

Create a LabVIEW program that does the following.

1. Calculates the strain at the location of the installed strain gage on the aluminum bar mounted as a cantilever beam. Assume you will measure V_s and V_{amp} (and calculate ΔV_{amp}) with the DAQ attached to the amplifier (reference the manufacturer's manual). Recall that ΔV_{amp} is amplified ΔV_g. The amplification factor should be set (consider expected strains,

magnitude of V_g, and range of analog to digital conversion) and noted for the lab. The (VI) will need to calculate ΔV from ΔV_{amp}. Display and record strain, V_s and ΔV_{amp}.

(a) Consider the strain gage amplification factor.

(b) Consider the geometry of the beam: cross-section is expected to be consistent, along with the moment arm of the applied load.

(c) The gage factor for most gages typically used is $2.1 \pm 0.5\%$.

2. Calculates, displays, and records the load applied by a weight placed on the beam in the fixture using the strain gage data. You will have to measure/estimate the location of the application of this load relative to the strain gage. The stress at the gage location can be an intermediate calculation, display, and record this result also.

3. Calculates, displays, and stores the load applied by a weight placed on the beam in the fixture using a calibration constant developed considering the strain gage ΔV_{amp} as a proportional transducer signal. The calibration constant will be developed in lab. Use the shifted and tared ΔV_{amp} signal as the transducer signal to simplify the process.

4. Includes the ability to modify the acquisition time and the sampling rate in your Start with using a 0.1 s acquisition time and a 1,000 Hz sample rate.

5. Includes the ability to estimate the standard deviation of V_{amp} and V_s, and store them.

6. Hints: Use a Gain SubVI in a while loop. Write data out with a write to spreadsheet tool. Wire channel AI0 as V_s and AI1 as V_{amp}. You will need a shift register/build array for each value you wish to write out. If you wish to plot any values vs. time, you will need to create a time array (Lab1). Use the plot indicator from the bottom of the Gain SubVI for a quick plot; remember to bundle two single-dimensional arrays to connect to the plot. Make any values that may need adjusting controls on the front panel (like moment arms). Put indicators on the raw data, channel AI0 and AI1.

2.9.6   EXPERIMENTAL TASK

1. Week 1, bond a strain gage on an aluminum beam. Follow Vishay's strain gage installation instructions or other provided installation instructions. Wire the gage in the three wire configuration.

2. Week 2, place marks on the beam to aid in accurately positioning the calibration weights as well as the soda can. Measure the diameter of the beverage can, and place a line across the top surface of the beam at the radius of the beverage can from the free-end on the beam. Keep notes on the precise geometry you use.

3. Mount your instrumented cantilever beam, as shown in Fig. 2.14. The beam is installed by loosening the screws in the support structure shown in Fig. 2.13 and gently sliding the rear of the beam into the constraining slot and carefully re-tightening the screws. Be careful not to damage your strain gage or wiring during this step.

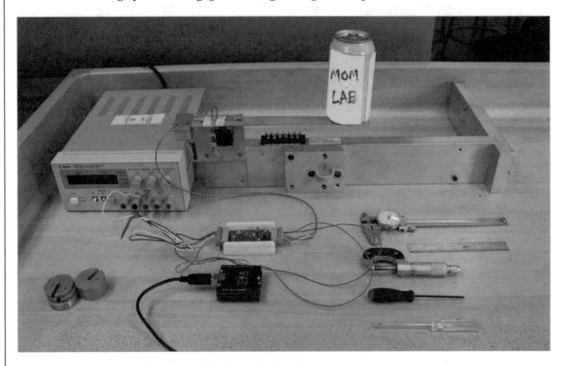

Figure 2.14: Assembled cantilever beam and the relevant wire connections.

4. Connect the three wires from your strain gage on the cantilever beam to the appropriate terminals on the strain gage amplifier. If a Tacuna® Strain Gage Amplifier is used, follow the wiring instructions for a three wire quarter bridge strain gage located in the amplifier manual. These details are shown in Fig. 2.15.

5. Plug-in or turn on power supply as required to power the strain gage amplifier.

6. Set up the amplifier. If using the Tacuna® Strain Gage Amplifier, set the zero screw so that the unloaded voltage from the amplifier is approximately 2.5 VDC. Chose an appropriate value for the gain window (on the Gain.vi) and report it in the procedure section of your lab write-up. Tare your VI by subtracting the V_{amp} measured when the beam is unloaded from subsequent measured V_{amp}.

7. Calibrate: Weigh the calibration weights on the commercial scale. Use the 50 g, 100 g, and 200 g weights. Take data using your instrumented beam for "50 g," "100 g," "200 g,"

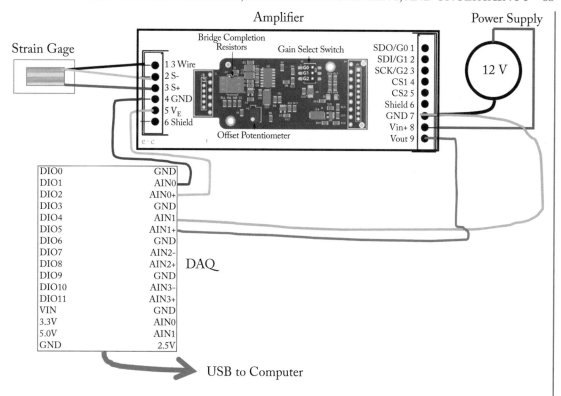

Figure 2.15: Magnified view of cantilever support, DAQ, and amplifier connections.

"300 g," and "350 g" cases. Use the weights measured with the commercial scale, not the values stamped on them (Why?). Place the scale-measured weights and their respective ΔV_{amp} into a spreadsheet and fit a linear function (a "Trend line") that allows you to determine a weight from a given ΔV_{amp}. *(You can use Excel, LabVIEW, Matlab, or any other program you wish to use.)* The slope of the trend line is the calibration constant, if the independent variable is the ΔV_{amp}. Note: if the calibration constant is developed using the raw V_{amp}, then the y-intercept must also be included in the calibration procedure (and uncertainty estimates).

8. Weigh the unopened can with your scale ten times. Each time, remove and place the can on the beam at the same location. We are examining the repeatability of placement. Weigh the can with the COTS scale once for comparison.

9. Open the can and take a gulp: ensure the VI records weight estimated from both processes (calibration and strain measurement) using your beam scale. Repeat gulp and weight until

empty. Allow the VI to run a minimum of 5 iterations for each gulp (if sample time set to 1 s, 5 s), and export to excel when the can is empty.

10. Weigh the empty can on your scale and on the commercial scale.

11. Chose an object from your pocket, weigh and also weigh on the commercial scale. Record weights Strain-based, calibration-based, and commercial scale reading in grams).

12. When finished weighing, unplug power supply.

2.9.7   ISSUES TO BE DISCUSSED IN THE LAB REPORT

1. Estimate the maximum weight that can be measured by the cantilever beam if the material is 6061 T6 aluminum. Take care to note that your beverage can weight is centered at a distance away from the end of the beam. This calculation is necessary to prevent using large weights which may otherwise plastically deform the beam or exceed the capacity of strain gage. Assume that the gage is not a limiting factor.

2. Report the applied weight of the full can found with two ways: (1) from the measured voltage ΔV_{amp} and your calibration curve and (2) from the computed stress. (This is a cantilever beam for which you should be able to compute the surface stress as a function of endpoint load and, conversely, the load from the stress.)

3. Estimate the uncertainty of both weight measurements of the full beverage. For the first method, consider the uncertainty in the commercial scale (calibration weights, Y), and the uncertainty in the ΔV_{amp} (calibration voltages, X). Assume that these uncertainties are reflected in the slope that you find in the linear fit. Propagate accordingly. For the second method, consider the uncertainty in the voltage measurements, material modulus, gage factor, and relevant measured lengths. Compare and discuss these two uncertainties.

4. Calculate and report the mean and standard deviation of the weight found from the placement of the full can ten times based on one of the weight measurement methods. Select AND NOTE, the method based on a quantitative argument developed in discussion issue 3.

5. Use the Student's t-distribution, 95% confidence interval, to estimate the statistical uncertainty of the weight of the unopened beverage as it is placed on the beam ten times. Report and discuss in light of the uncertainties found in Step #3. Is repeatability of can placement an issue? Use the results of this step to discuss the effect the placement might have on the overall uncertainty.

6. Determine and report the mean and standard deviation of your gulp size based on one of the weight measurement methods. Select the method based on a quantitative argument concerning uncertainty developed in discussion issue #3.

7. Discuss any variation found in determining the weight of the empty can with the beam scale and the commercial scale.

8. Recommend improvements to reduce the uncertainty for the instrumented cantilever beam.

9. Does the weight of the beam affect the calibration? Explain.

2.9.8   EQUIPMENT REQUIREMENTS AND SOURCING

- Multi-function DAQ with USB cable: Minimum 2 channel 14 bit ADC, USB interface, LabVIEW support. Typical vendor: Out of The Box SADI DAQ,

  https://ootbrobotics.com/.

- Aluminum bar, 1″ wide x 1/8″ thick x ~12″ long, Typically 6061 T6, typical supplier: McMaster Carr.

- Student Strain Gage, 120 Ω, Typical Part: Micro-Measurements CAE-13-240UZ-120.

- Strain gage amplifier, typical part: Tacuna Systems EMBSGB200_2_3 configured for 1/4-bridge.

- Power supply: 6-12 VDC, source: various.

- Calibration weights: slotted brass weights. Typical vendor: Manson Labs, others.

- Commercial scale, typical part: Symmetry EC2000 or similar.

- Measurement tools in lab (rulers, micrometers, micrometers, etc.).

2.10   APPENDIX: MONTE CARLO SIMULATION TO ESTIMATE UNCERTAINTY IN A LINEAR FIT

The calibration method used in Lab 2 generates a proportionality constant between a force applied on the beam and the voltage measured by a DAQ. Figure 2.9 shows an illustration of the utilization of a M-C simulation to examine the possible variation of the proportionality constant as a function of the uncertainty in the calibration weights and measured ΔV_{amp}. This approach is based on simulating a new set of data points from the original measured points by considering the random error attributed to each measurement. A new simulated measurement is found by randomly perturbing the original measurement within the range of the random uncertainty in the original measurement. It is assumed that the random error is Gaussian in nature and that the uncertainty found represents one standard deviation.

A spreadsheet program is useful for implementing this simulation. Columns are set up to represent each measurement, either weight or voltage. Rows represent each data set, and a linear fit is performed on each row of simulated data. The resulting slope is stored in a column. The average and the standard deviation can be developed for the column of slopes, with the standard deviation representing an estimate of the uncertainty in the slope as a result of the uncertainty in the measured data points. The data necessary to perform the simulation includes the raw measured weights and voltages, and the uncertainties in each measurement. As noted, simulation of a new set of data should address the distribution of the random uncertainty.

The following excel code randomly generates a simulated measurement from a set that is distributed in a Gaussian manner about the measured data with a standard deviation equal to the stated random uncertainty in the measurement:

=NORM.INV(RAND() (cell address of measured data, cell address of measured data uncertainty).

Use this Excel function to generate an estimated value of V_{amp} or weight. After all Vamp's and weights for a data set are generated, use the following Excel code in a cell to generate the slope of the simulated data set:

=SLOPE(independent variable range, dependent variable range).

It is left to the reader to determine the standard deviation of the resultant column of simulated slopes. The M-C simulation is open ended in that the number of simulations needed is not known. It is appropriate to increase the number of simulated data point sets until the standard deviation does not change with respect to the resolution of the data. The typical number of simulations needed for this problem are in the thousands or tens of thousands.

L A B O R A T O R Y   3

Stress–Strain Response of Materials

PART A: THEORY

3.1   INTRODUCTION

Structural components in engineering applications are subjected to complex loads and under-standing the influence of these loads is crucial to their safe design. However, because of the complexity of operating environment, we try to understand the material behavior in simplified loading conditions and then formulate failure theories for deformation behavior under complex combination of loads. To this end, we focus on material deformation under simple uniaxial (1D) tension and compression. Let us review two fundamental concepts in mechanics of materials.

(i) *Stress:* Intensity of internal force at a given point. The intensity is measured as, internal force (P) divided by area (A) at that location. Note that P is not the external applied force. The internal force is the resultant of all the applied external forces (and moments) acting at that point on a given cross sectional plane. To obtain resultant internal force, a cut section is made at that point and the forces are vectorially summed. For simple uniaxial loading along x-axis on a bar, the stress acting on a plane perpendicular to the loading axis is given by

$$\sigma_x = P/A \tag{3.1}$$

(ii) *Strain:* Change in length (dl)/original length (l). If the change in length is in axial direction x, then

$$\varepsilon_x = dl/l \tag{3.2}$$

In both of the above equations, the denominators A and l refer to either original dimensions or current dimensions. If original dimensions are used, then they are called "engineering stress" and "engineering strain," and if current dimensions are used, they are referred to as "true stress" and "true strain."

To conduct tension test (see Fig. 3.1), we typically use a dog bone-shaped specimen, grip it on two ends, and pull on it. The gage section is assumed to undergo uniform and uniaxial deformation and this deformation (elongation) is measured using an extensometer. From the measured load and elongation of the specimen, we can plot engineering stress and engineering strain. The word "engineering" is used here to indicate that the stress and strain are calculated based on the original cross-sectional area (A) and original gage length (l), respectively, of the specimen. ASTM recommended specimen dimensions are often used by the industry. However, we use rectangular-shaped specimens in this lab.

Typical stress–strain response of metals which undergo elastic-plastic deformation is shown in Fig. 3.1. The initial response is linearly elastic with slope defined by the Young's modulus (E). Upon reaching the end of elastic response, metals yield (σ_y) and undergo plastic deformation where the strain increases at a faster rate than elastic response. During the plastic deformation the metal work hardens (increase in stress with an increase in plastic strain) and reaches a maximum value (σ_{max}). This stress is the "ultimate strength" of the material. Until this

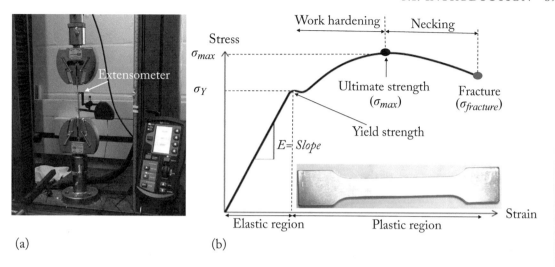

(a)   (b)

Figure 3.1: (a) Typical metallic dog bone-shaped specimen with extensometer to measure elongation of the gage section during tensile loading and (b) and its tensile stress–strain response.

point, the strain in the gage section is uniform. Beyond this point, the specimen starts to neck in the gage section, i.e., the deformation localizes at one cross section (not uniform in the entire gage section anymore). The location of this localized strain is random, due to random nature of microstructural defects present in the material which trigger this localization. With continued application of tensile load, the diameter (or cross-sectional area) of the specimen in this location gradually decreases. The stress starts to fall down and eventually the specimen fractures. The associated stress is labelled $\sigma_{fracture}$. The specimen fractures at 45° to the loading axis. This is the plane of maximum shear stress which is given by $\tau_{\max} = \frac{\sigma_{\max}}{2}$. Note that, although the plot in Fig. 3.1 reveals stress drop after reaching the ultimate stress, in reality, the stress in the material continues to raise because of the continuous decrease in the cross sectional area of the specimen at the fracture location. This increase in stress will be reflected if we plot instantaneous area (not the original area) in the calculation of stress, which will be called "true-stress." Similarly, if we use current length as opposed to original length, we can also calculate "true-strain" in the specimen. However, we have defined engineering stress and strain (Eqs. (3.1) and (3.2)) based on original cross-sectional area and original gage length and hence the decrease in stress beyond the maximum stress is seen in the plot.

The linear portion of the stress–strain curve has a slope E, defined as Young's modulus. In this regime, the stress and strain are related by Hooke's law, defined as

$$\sigma_x = E\varepsilon_x \tag{3.3}$$

In 1D loading, the axial deformation along the length of the specimen also causes change in lateral dimensions. We define the lateral strain (ε_{lat}) as change in diameter (dD) over original diameter (D), i.e.,

$$\varepsilon_{lat} = dD/D \tag{3.4}$$

The axial strain and the lateral strain are related by,

$$\text{Poisson's ratio} \quad \nu = -\varepsilon_{lat}/\varepsilon_{ax} \tag{3.5}$$

Here $\varepsilon_{ax} = \varepsilon_x$. The negative sign in the above definition indicates that the change (decrease) in lateral dimensions is in the opposite direction of the axial change (increase in length) in dimensions which is taken as positive. Knowing E, ν, and axial strain, we can calculate the change in diameter or area of the specimen under uniaxial loading.

Question:

1. Why do we plot stress (σ) vs. strain (ε), and not load (P) vs. elongation (ΔL)?

   Load and displacement depend on the size (geometry) or dimensions of the specimen. However, our aim is to define a response that is unique to each material (i.e., steel, aluminum, composite, ceramic, etc.). By normalizing the load (force) and displacement (elongation) with the cross-sectional area and original length, respectively, we can obtain the response that is unique to each material, irrespective of its size.

   In this laboratory, we intend to study material response to 1D (uniaxial) loading under tension and compression. The stress–strain response of a material can be obtained either by controlling applied load or controlling the displacement of the cross-head of the machine. Accordingly, we define the stress–strain responses as either load controlled or displacement controlled.

3.2   TENSILE STRESS–STRAIN RESPONSE OF MATERIALS

3.2.1   LOAD-BASED STRESS–STRAIN CURVE

Here we use a thin copper wire to determine its load-based stress–strain curve. In this lab, a linear variable differential transducer (LVDT) is used to measure the elongation of the wire when selected weights are placed on its loading platform (or hanger). LVDT converts displacement into voltage. It consists of a plastic hanger with a metal rod at its end which moves inside the core of a copper coil wound in a specific fashion. Initially, in the no-load condition, the metal rod is exactly at the center of the coil. When weights are placed on the hanger (attached at the end of the wire), the resulting elongation of the wire (displacement) and hence the metal rod moves inside the copper coil thus causing a change in voltage which is measured during the experiment. It is important to understand the operation of LVDT which consists of three solenoid coils placed end-to-end around a tube. The central part of the coil is called "primary (P)" and the outer coils are named "secondary (S)." A thin conducting wire is wound in a one-orientation in the primary section, and in the opposite orientation in the secondary sections. Also, the two secondary sections (S_1 and S_2) are connected electrically such that the output is the differential voltage between them (and hence the name "differential" in LVDT). The primary coil is actuated by alternate current (AC), and this induces equal voltages in the two secondary coils when the rod is exactly at the center of the cylinder. However, the output voltage of LVDT will be zero because the output is the "differential" voltage between the two secondary coils. When a metal rod (also referred to as core) moves inside this coil (due to elongation of the copper wire when load is placed on the hanger), depending on its position with respect to the central coil, a differential voltage is generated.

When the metal rod (also referred to as core) is asymmetrically positioned with respect to the P-coil, for example, and if the rod moves up, higher voltage is generated in the upper S-coil than the lower S-coil and vice versa. When the rod is situated exactly symmetric (i.e., the metal core is centered), the generated voltage in S_1 and S_2 are equal, i.e., $E_1 = E_2$ and $\Delta E = E_2 - E_1 = 0$. When the core moves towards the bottom, voltage in S_2 is greater than S_1, i.e., $E_2 > E_1$, $\Delta E > 0$ and when the core moves towards the top, the voltage in S_1 is greater than S_2, i.e., $E_2 < E_1$, $\Delta E < 0$. These three scenarios are illustrated in Fig. 3.2.

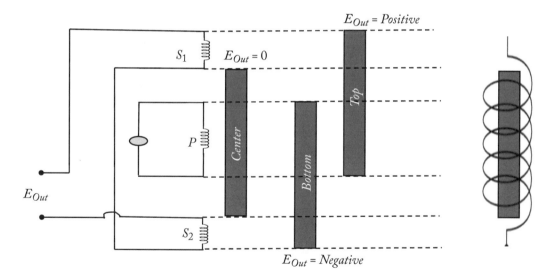

Figure 3.2: Schematic of the primary and secondary coil arrangement in a LVDT and illustration of electromagnetic force evolution when a metal rod moves through the coil.

Thus, the phase of the output voltage and its amplitude are related to the direction of motion (up or down displacement) and magnitude of displacement, respectively, of the metal rod. In general, the coils are designed to provide a linear output over a given displacement range, and hence the term "linear variable (LV)" in LVDT.

If the LVDT used in the lab has a voltage range ΔV, typically on the order of few volts (e.g., ± 5 V) for a specified displacement ΔL (e.g., 1 in or 25.4 mm), the constant between voltage generated and elongation can be determined as:

$$\Delta L = C(\Delta V) \quad \text{i.e.,} \quad C = \frac{\Delta L}{\Delta V}\left(\frac{\text{mm}}{\text{Volt}}\right) \tag{3.6}$$

The constant C is usually supplied by the manufacturer.

The experimental setup to obtain a load-based stress–strain response using an LVDT is shown in Fig. 3.3. The thin wire is wrapped around a pulley and a known weight is placed on the hanger connected to LVDT. The resulting voltage ΔV (or ΔE) is measured. Knowing the value of C, one can calculate ΔL. From given weight and cross-sectional area of the wire, we can determine stress $\left(\frac{F}{A}\right)$ and from known ΔL and length of the wire we can calculate the strain $\left(\frac{\Delta L}{L}\right)$. Repeat the process for different weights and plot the stress–strain curve for the wire.

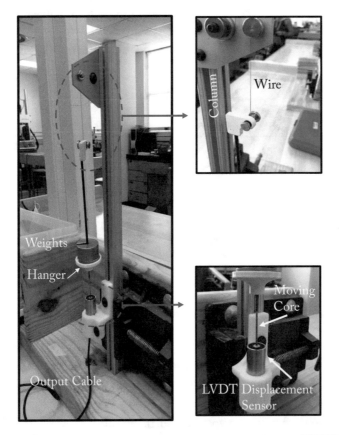

Figure 3.3: Load-based stress–strain response test equipment using an LVDT.

One may need to be cautious about the weights being used to calculate stress. Note that the wire is already in tension due to the initial weight of the hanger even before the first weight (W_1) is placed on the hanger. This elongation was never measured at the beginning of the experiment because of the need to have the hanger in place to get the LVDT ready for measurements. So there is small error in your calculation of stress because the weight of the hanger is not accounted for in each weight placed on the hanger. Since the stress–strain curve of the wire is yet unknown, one can estimate the elongation due to this hanger by weighing the hanger separately and then adjusting the weights accordingly or using the load-elongation (or stress–strain) curve to estimate the elongation for this weight.

3.2.2   DISPLACEMENT-BASED STRESS–STRAIN CURVE

This test is performed using a servo-controlled machine. In majority of situations, displacement-based tensile testing is preferred because the maximum displacement required to cause failure of a tensile test specimen is small and the machine can be programmed to move only by this limited amount without exceeding its capacity. Besides, under load-control, when a specimen fails, the controller continues to move the loading column until the set load is reached. But, because the specimen failed, there is no load to be applied and hence the controller instructs the loading column to move and achieve the set load, potentially jamming into other parts of the machine and casing damage. This behavior is similar to tug-of-war between two teams. When the rope breaks, the two teams suddenly lose control and fall back. This is because, at the instant of rope failure, the force is still being applied and upon breakage of the rope, the force (or the stored energy) still exist and is now taken up by the fall. In the case of the machine, it has been instructed to apply the force, but the specimen has failed and so it continues to move to apply the load, and can potentially jam in to other components unless it is instructed immediately to stop. Due to these reasons, displacement control is preferred over load control while conducting these tensile tests.

In this laboratory, the uniaxial stress–strain response of several materials (a ductile metal, a composite, a plastic, and a ceramic) are determined in either tension or compression using appropriate test geometries shown in Fig. 3.4. While the first three materials are loaded in tension, the ceramic (or a brittle material) is tested in compression.

To measure small elongation of the specimen (e.g., metal) LVDT or strain gage or extensometer may be used, but for large elongations (e.g., plastic or polymer tensile testing) the LVDT or strain gage-based extensometer range is exceeded, and so the actual displacement of the loading column is recorded for calculation of strain. In this laboratory, an Instron®machine shown in Fig. 3.4b may be used, where the load is measured by a load cell situated in the upper cross head and the displacement is measured using a strain gage-based extensometer. As a side note, students should note that some extensometers have a cantilever beam with four strain gages (two on the bottom and two on the top surface) arranged in full bridge-configuration, as discussed in Laboratory 2.

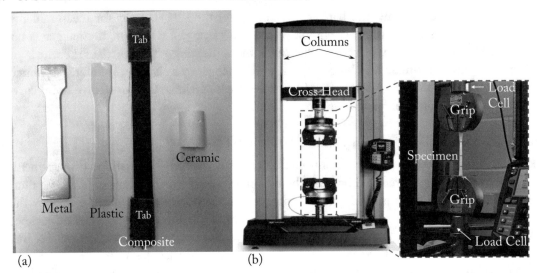

Figure 3.4: (a) Specimen geometries for tension and compression testing and (b) test machine to conduct uniaxial tensile and compression tests.

3.2.3   TENSILE RESPONSE OF MATERIALS

Metals

The typical tensile stress–strain response of a metal was described in Fig. 3.1. Let us now discuss the failure behavior of a metal under tensile loading. Plastic deformation in ductile metals is driven by shear stress and identification of the amplitude and orientation of the maximum shear stress plane is important to determine the impending fracture location and fracture growth direction. Obviously, for 1D uniaxial loading, it is trivial, i.e., the fracture plane is at 45° to the loading axis (plane of maximum shear). However, in complex stress states, one needs to calculate principal stresses and maximum shear stress as well as their orientations either using stress-transformation equations or Mohr's circle (MC). For our uniaxial tension test, the following situation can be visualized, as depicted in Fig. 3.5.

Note that the stresses at the indicated point in Fig. 3.5a are shown on Mohr's circle in Fig. 3.5b, and recall from the Mechanics of Materials class that the maximum shear stress plane on the Mohr's circle is 90° from the principal planes which lie on x-axis. These shear stress planes are at 45° to the principal planes (horizontal and vertical planes) on the specimen. The failure plane of the specimen is shown in the inset of Fig. 3.5a which is at 45° to the loading axis.

Determine the unknown material given its stress–strain curve: While conducting tensile test in the laboratory on a metallic material, you have not been told what material it is. You are only given the stress–strain curve (determined above) and asked to determine which material it

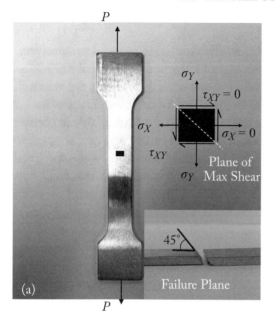

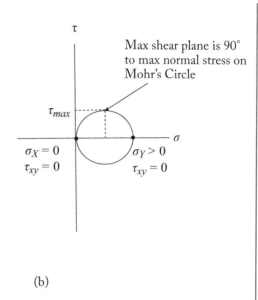

Figure 3.5: (a) Stress at a point along a tensile fracture plane of the test specimen and (b) Mohr's circle (MC) indicating the stresses on horizontal and vertical planes at any point in the gage section of the specimen. The failure plane is indicated in the inset of (a).

is. To identify a material (for example aluminum or alloy type), carefully measure the Young's material E, yield stress σ_y, and fracture stress σ_f of the material from the stress–strain curve. It also helps to assess the uncertainty in these values (from other lab groups). Look at heat treatments of these metals from literature values (on the web) to further narrow down the material type.

Brittle Materials (Ceramics, Rocks, Concrete, etc.)
Unlike metals, ceramics do not exhibit plasticity. They are brittle and hence fail catastrophically once cracks are initiated. Ceramics are also weaker in tension than in compression. Therefore, they are often used as load bearing elements in compression. These materials exhibit elastic response initially and then fail suddenly upon reaching a maximum stress value. Since there is no macroscopic plastic response, the stress–strain curve is a straight line until fracture. The failure is governed by tensile stress (not shear stress as in metals) because tension is needed to open a crack and cause fracture.

Since ceramics are hard and break easily, it is difficult to machine them into dog bone-shaped specimens for tension tests. Knowing that ceramics have only elastic response, and that we are looking for uniaxial response of materials, it is easier to make these brittle ceramics into regular cylinders and test them under compression (see Fig. 3.4a).

Can you now guess the failure plane for a ceramic when loaded in uniaxial compression? Where is the maximum tension in the specimen when loaded in compression? Well, the easiest way to find out is to draw the Mohr's circle. The relevant stress-state and Mohr's circle as well as cracks initiated in the ceramic specimen are shown in Fig. 3.6. Notice that cracks are parallel to the loading axis (vertical cracks) and grow along the length of the cylinder. This is counter intuitive! For any crack to form, a tensile state must exist. Tension causes two new surfaces to form by separating the material. Theoretically, there is no tension at any point in the specimen (see Mohr's circle). But the lowest magnitude of compressive stress is zero and this stress acts on a plane parallel to the loading axis as seen in the Mohr's circle. Therefore, potential for crack initiation is highest here. So, why does failure occur on a plane with zero tensile stress? This is because ceramics have surface defects where stress concentration persists. When a specimen is compressed, the ceramic specimen expands laterally, and the resulting hoop stress causes tensile cracks to initiate from the preexisting defects on the surface. These cracks grow parallel to the loading axis and cause failure of the specimen into columnar fragments as seen on the test specimen.

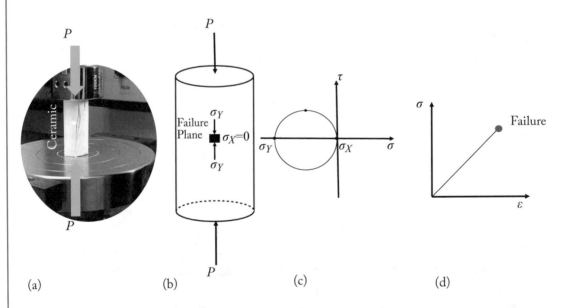

Figure 3.6: (a) Compression testing of a brittle material where axial cracks develop at peak load, (b) free body diagram of the specimen with a point along the fracture plane, (c) Mohr's circle representation of the stress state at any point on the surface, and (d) tensile stress–strain curve for a typical brittle specimen.

Fiber-reinforced Composites

We will focus on a uniaxial carbon fiber-reinforced composite with epoxy matrix. The carbon fibers are aligned parallel to the length of the specimen; see Fig. 3.4 and the loading direction is parallel to these fibers. To avoid stress concentration at the loading grips, the specimen is bonded with metal tabs on top and bottom where the specimen is to be held during testing (see Fig. 3.4a). When pulled in tension, the composite fails first in the brittle matrix phase and then eventually the fibers start to break (see Fig. 3.7). The fibers are brittle and snap when failure load is reached, and this failure process can be noted by audible sound during testing. As fibers break, the load suddenly drops, and the remaining intact fibers take up the load. Because the specimen is continuously being pulled, the stress continues to be maintained and new set of fibers break. The load drops again and the process continues until the specimen is fully separated. By carefully examining the failure plane the student is urged to discuss the orientation of the failure plane with respect to the loading axis. Recall that the fibers are brittle and the loading type is tension. So, what is the orientation of failure plane when fiber breaks?

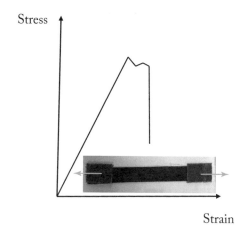

Figure 3.7: Stress–strain curve for fiber-reinforced composite. Early-stage fiber breakage is seen in a single-fiber strand.

Plastics

Here, we use a thermoplastic nylon polymer. This polymer with some reinforcements is often used in many automotive components and house hold fixtures. At the microstructure level, the thermoplastic consists of polymeric chains which can unfold when pulled in tension (similar to spaghetti) and hence they undergo large plastic strains, well in excess of 100%. The stress to cause plastic deformation is small compared to metals and composites discussed before. The stress–strain curve for a polymer and the failure modes observed in various materials used in this laboratory are shown in Fig. 3.8.

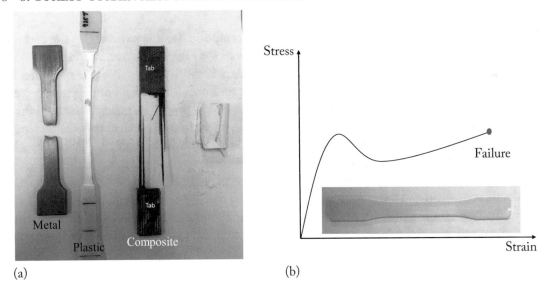

(a)   (b)

Figure 3.8: (a) Failure modes in various materials and (b) stress–strain curve for polymers (soft plastic).

Discussion Points for Each Specimen

After obtaining the stress–strain curves for all materials, the following points should be discussed in the report.

1. Modulus of elasticity: Choose an approximately linear portion of the stress–strain curve. Fit a "linear" trend line. Report stress range chosen.

2. Yield strength if discernable otherwise use 0.2% offset yield.

3. Ultimate strength.

4. Breaking strength (recall from experiment that plastics fail outside the extensometer range).

5. Percentage of the elongation to ultimate stress.

6. Toughness: Area under the stress–strain curve. Units: J/m^3 or the product of stress (σ) and strain (ε) $= \frac{N}{mm^2} \times \frac{mm}{mm} = \frac{N}{mm^2}$.

   Use numerical integration (e.g., trapezoidal rule) to calculate toughness.

   Area of the trapezoid $= \Delta\varepsilon \cdot \sigma_{avg} = (\varepsilon_1 - \varepsilon_2) \cdot \left(\frac{\sigma_1 + \sigma_2}{2}\right)$.

   Question: Can you include the area under the stress–strain curve after the σ_{UTS} for toughness calculation?

7. When comparing materials for their strength, it is beneficial to compare the strength per unit weight (or density). This measure is called "specific strength." It allows an engineer to make judgments based not only on the strength of a material but also its density which contributes to the total weight of the structure.

3.3 UNCERTAINTY IN STRESS, STRAIN, AND ELASTIC MODULUS

While reporting failure stress and strain, one must always indicate the uncertainty in these values.

3.3.1 UNCERTAINTY IN STRESS
By definition stress,

$$\sigma = \frac{F}{A} \tag{3.7}$$

where $A = $ thickness $(t) \times$ width (w).

The uncertainty in stress is given by

$$U_\sigma = f(U_F, U_A) = (U_F, U_t, U_w) \tag{3.8}$$

Uncertainty in load cell measurement, U_F, can be obtained from load cell manual (supplied by the manufacturer). For the machine used in the laboratory find the maximum load from the manual or from the website of the manufacturer (e.g., Intron 5567; $F_{max} = 30$ KN).

Load measurement accuracy is generally given for different ranges as:

$$\text{If} \quad F \geq \frac{1}{100} F_{max} \to 0.4\% \quad \text{uncertainty}$$

$$\text{If} \quad \frac{1}{250} F_{max} \leq F \leq \frac{1}{100} F \to 0.5\% \quad \text{uncertainty}$$

To simplify the problem, we may assume 0.5% uncertainty in force measurement (U_F) and neglect ranges (we have selected higher uncertainty values).

Uncertainty in stress can now be calculated as

$$U_\sigma = \sqrt{\left(\frac{\partial \sigma}{\partial F}\right)^2 (U_F)^2 + \left(\frac{\partial \sigma}{\partial A}\right)^2 (U_A)^2} \tag{3.9}$$

$$U_\sigma = \sqrt{\left(\frac{1}{A}\right)^2 (U_F)^2 + \left(\frac{-F}{A^2}\right)^2 (U_A)^2} \tag{3.10}$$

Since $A = wt$, U_A can be calculated using similar calculation. Note that U_t and U_w depend on the type of scale used (micrometer or caliper). It is common practice to use half the resolution of the instrument as the uncertainty for that measurement.

3.3.2   UNCERTAINTY IN STRAIN U_ε

This value is usually obtained from the extensometer manual. Usually, the error in strain should not exceed $\pm 0.5\%$ of strain ($U_\varepsilon = \pm 0.5\%$).

3.3.3   UNCERTAINTY IN ELASTIC MODULUS (MONTE CARLO SIMULATIONS)

Report uncertainty in "E" only for the metal specimen using the M-C simulations (discussed in Section 2.6.1).

PART B: EXPERIMENT

3.4   LOAD CONTROLLED TENSILE TESTING OF A METALLIC WIRE

3.4.1   OBJECTIVE

This lab is designed to give experience in developing stress–strain diagrams based on a load controlled test. In this lab you will use COTS weights to incrementally load a wire and measure elongation with a LVDT. Throughout the process, you are to monitor uncertainty and develop estimates of the accuracy and precision of your results.

3.4.2   BEFORE LAB

1. Review the MOM relationships associated with a wire under tensile loading: deflection, stress, strain, ductile behavior, and offset yield point.

2. Review load controlled tensile testing. Can fracture strength be detected?

3. Review the use of a micrometer.

4. Familiarize yourself with LVDT's. See class notes and:

   http://www.macrosensors.com/lvdt_tutorial.html,

   http://www.rdpe.com/us/hiw-lvdtdc.htm

3.4.3   PRELAB QUESTIONS

1. What is the expected load needed to yield the 32 AWG copper wire?

2. What is the expected load needed to fail the 32 AWG coper wire?

3. If the LVDT has 20 mm of travel, what is the maximum length of 32 AWG wire appropriate to ensure the wire will break before the LVDT saturates?

3.4.4   EQUIPMENT AND SUPPLIES NEEDED

See Fig. 3.9.

- Laptop computer with LabVIEW installed and functioning.

- Multi-function DAQ.

- Wire.

- LDVT.

- Brass weights.

- Tensile loading fixture.

- Rulers, weighing scales, dial caliper, and micrometer.

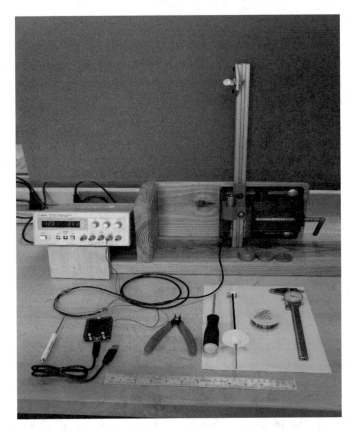

Figure 3.9: Equipment and supplies needed for conducting experiment.

3.4.5   PROBLEM STATEMENT

Examine the load-displacement properties of a wire. Use the information to develop a stress–strain diagram for the wire. Identify the modulus of elasticity, elastic limit, yield point, 0.2% offset yield point, and ultimate strength. Develop uncertainties for each identified property.

3.4.6   REQUIRED LABVIEW PROGRAM (VI)

Create a LabVIEW program to accomplish the following.

1. Measure and store the LVDT output voltage.

2. Develop the standard deviation for each measured LVDT voltage and store (see Laboratory 1).

3. Tare and apply a calibration constant to the LVDT-measured voltage to arrive at a change in displacement. Pay attention to units.

4. Allow the input of the initial length of the wire.

5. Calculate, display, and log the strain in the wire continuously.

6. Allow the input of the loading weight values.

7. Calculate, display, and log the tensile stress the wire is undergoing as it is loaded.

8. Include the ability to modify the acquisition time and the sampling rate. Start out using a 0.1 s acquisition time and a 1,000 Hz sample rate.

9. Hints: Use a Gain_xxxx SubVI in a while loop. Write data out with a write spreadsheet tool. Wire channel AI0 as the LVDT voltage. You will need a shift register/build array for each value you wish to write out. If you wish to plot any values vs. time, you will need to create a time array. Use the plot indicator from the bottom of the Gain_xxxx SubVI for a quick plot, remember to bundle two single-dimensional arrays to connect to the plot. Make any values that may need adjusting controls on the front panel. Put indicators on the raw data, channel AI0. Use the data array and an index array block to extract the raw data from the DAQ for "b".

   Note: there is no need to develop the stress–strain curve in the VI. It may be done using the stored data after the lab.

3.4.7   CONNECTIONS REQUIRED

Connect the LVDT to your DAQ following the wiring instructions for the LVDT. Select the appropriate volt range on the DAQ.

3.4.8   EXPERIMENTAL TASK

Be sure to log appropriate loads with appropriate forces.

1. Power on the LVDT. Check that you have data from the LVDT in your VI.

2. Input the calibration constant for the LVDT.

3. Cut a piece of wire between 20 and 25 cm long.

4. Attach one end of the wire to the weight hanger, as pictured in Fig. 3.10. Bind one end of the wire under the thumb screw.

Figure 3.10: Illustration of wire attachment to hanger.

5. Attach the other end of the wire to the top of the tensile loading fixture, as seen in Fig. 3.11. Clamp the wire between washers between the thumbscrew and the aluminum plate. The wire should pass between the two binding posts.

6. Loosen the LVDT carrier from the frame and raise it, inserting the LVDT core that is on the bottom of the weight hanger into the LVDT cylindrical body. Adjust the LVDT carrier so that the VI reads an LVDT voltage at the bottom of the device's range (Fig. 3.12a), allowing for full utilization of the LVDT's travel as the wire stretches. The fully assembled test fixture is shown in Fig. 3.12b.

7. Characterize the load-deflection characteristics of the wire.

   (a) Incrementally load the wire and take data up to failure. You can check for elastic/plastic behavior by loading and then unloading, and examining the resultant deflections (look for permanent deformation).

   (b) Evaluate your initial data to establish likely regions of the elastic limit, the yield point, and the ultimate strength.

   (c) Iterate the process with new wire as many times as needed to identify these material characteristics.

8. Issues to check for.

   (a) The LVDT is only linear between −5 V DC and +5 V DC. If the experiment is started outside of this range, the results will be incorrect.

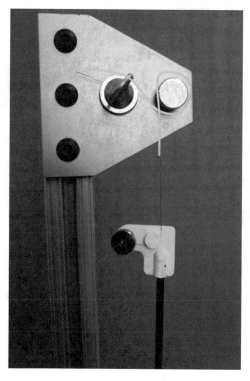

Figure 3.11: Illustration of wire attachment to top of loading fixture.

(b) Check the LVDT's linear range of travel (~20 mm). If the length of wire used allows elongation greater than 20 mm before ultimate load is reached, the measurements will be incorrect.

(c) Some wires have significant creep. Take length measurements as soon as weight is placed. Do not wait for long and allow creep deformation of wire.

9. Examine the fracture surfaces under an optical microscope.

3.4.9   ISSUES TO BE DISCUSSED IN THE LAB REPORT

1. Develop and present in the report a stress–strain diagram for the material tested.

2. Develop and present in the report an estimate of the uncertainty in the strain.

3. Develop and present in the report an estimate of the uncertainty in the stress.

4. Report the Modulus of Elasticity, Elastic Limit, Yield Point, and Ultimate Strength. Include the appropriate uncertainties.

(a)

(b)

Figure 3.12: (a) Illustration of positioning of LVDT core into the LVDT cylinder and (b) fully assembled test apparatus.

5. Utilize Table 3.1 of properties for your analysis and discussion.

Table 3.1: Copper properties

Material	Copper
Finish	Bright
Bendability	Bend and stay wire (soft temper)
Specifications	ASTM B3
Tensile Strength	200 MPa (29000 psi)
Wire diameter	0.2 ± 0.00025 mm (0.008 ± 0.0001 in)

3.4.10 PRINCIPAL EQUIPMENT REQUIREMENTS AND SOURCING

- Multi-function DAQ with USB cable: Minimum 2 channel 14 bit ADC, USB interface, LabVIEW support. Typical vendor: Out of The Box SADI DAQ,

  `https://ootbrobotics.com/`

- LVDT: RDPE.com, DCTH400.

- 32 AWG copper wire, Vendor: McMaster Carr.

- Calibration weights: slotted brass weights. Typical vendor: Manson Labs, others.

- Commercial scale, typical part: Symmetry EC2000 or similar.

- Measurement tools in lab (rulers, micrometers, etc.). Various.

3.5 DISPLACEMENT-CONTROLLED TENSILE TESTING OF MATERIALS

3.5.1 OBJECTIVE

This lab is designed to give experience in developing stress–strain diagrams for various materials. In this lab, tensile tests and compressive tests of various materials will be performed using a displacement controlled Universal Testing Machine (UTM). The results of this lab will be presented as stress–strain diagrams in a lab report. Throughout the process, you are to monitor uncertainty and develop estimates of the accuracy and precision of your results.

3.5.2 BEFORE LAB

1. Review the use of a micrometer.

2. In this lab, there is only one UTM and the experiments will be performed by the lab instructor. During the test, data will be collected by the computer and the results will be displayed on the screen. There is no LabVIEW VI required for this lab.

3. Review: ductile behavior, plastic behavior, offset yield points.

3.5.3 EQUIPMENT AND RESOURCES NEEDED

See Fig. 3.13.

- Instron® 5967 Universal Testing Machine with 30 KN Load Cell.

- Instron® model 2630-116 Extensometer.

- Tension and compressive specimens.

- Calipers and micrometers.

- Optical/Digital microscope.

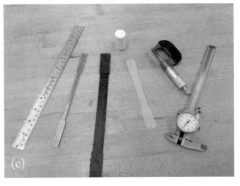

Figure 3.13: Equipment and specimens needed. (a) Instron® electro-mechanical test frame, (b) extensometer, and (c) specimens and measurement tools.

3.5.4   PROBLEM STATEMENT

Given an unknown metallic tensile specimen, a uniaxial carbon fiber tensile specimen, a nylon tensile specimen, and a plaster of Paris compressive specimen. Utilize the Instron® UTS to capture load-displacement data for each specimen. Develop a stress–strain curve for each of the test specimens from this data. One of the test specimens is an unknown metal. You are required to generate a match to referenced material properties in an attempt to identify the unknown metal, considering the uncertainty in the estimated material properties. The other test specimen data is to be utilized to generate various material properties to be compared and discuss observed elastic, plastic, and failure behaviors.

Example Problem

Draw a stress–strain diagram for a material with the following properties. Label each point with the corresponding letter:

(a) Modulus of elasticity = 69 GPa.

(b) Yield stress 290 MPa.

(c) 0.2% yield stress (construct).

(d) Ultimate strength 320 MPa.

(e) Breaking strength 275 MPa.

(f) Plastic strain at failure is 10%.

(g) Label the strain ranges where strain hardening takes place.

(h) Label the strain range where necking takes place.

(i) What type of metal could this sample be?

3.5.5   EXPERIMENTAL TASK

1. Measure the cross section of the unknown metal test specimen with the micrometer and calipers supplied. Each student is to perform a measurement. If fewer than 10 measurements are taken, students are to continue measuring until 10 measurements are recorded. Log the measurements in lab. Uncertainty in these measurements should be determined statistically, by using a 95% confidence interval (see notes from Laboratory 1).

2. Measure the remaining specimens (composite, polymer, and ceramic) to adequately define their geometry for use in discussing their response to the load applied by the UTS. The uncertainty in these measurements should be taken as the uncertainty in the measuring instruments.

3. Burnish marks on the face of the unknown metal specimen and any other ductile tensile specimen every inch in the narrow gage section. This set of marks is used to determine plastic extension at failure. One student per specimen is required to perform this task. The student that performs step #3 will perform step #7.

   In the following, steps 4–6 will be performed by the TA or lab instructor.

4. Mount a tensile specimen in the wedge grips of the universal testing machine.

5. Attach the extensometer to the specimen, centered about its length. The extensometer will be used for the ductile tensile specimens that deform less than 40%. The student will need to use the crosshead displacement for the carbon fiber and the nylon specimens as their elongations are significantly more than the limits of the linear range of LVDT.

6. Apply a constant cross-head displacement rate to the sample and collect strain (from the extensometer), displacement (from the cross-head displacement of the testing machine), and load from the load cell, until the sample is broken. The initial cross-head speed is 7 mm/m. This speed is constant for the metallic and carbon fiber specimens. For the nylon specimen, after the load has become consistent, the crosshead speed is to be increased to 100 mm/m until failure or end of travel.

7. For the tensile ductile specimens, reassemble the specimen and measure the change in length of two burnished marks that were originally 51 mm (2 in) apart with the necked section centered between the marks. Report the distance by writing it on the sample data sheet.

8. Observe and note the surface texture of the failed specimen (both the sides as well as the facture surfaces).

9. After the tensile specimens are tested, the machine will be configured for compressive tests. A brittle compressive sample will be tested.

10. Data will be collected and uploaded to the website for students to access that data and perform analysis.

3.5.6   ISSUES TO BE DISCUSSED IN THE LAB REPORT

1. Plot the stress–strain response for each specimen.

2. For each tensile specimen, determine the following.

   (a) Modulus of elasticity.

   (b) Yield strength (if discernable).

   (c) 0.2% offset yield strength (if needed).

   (d) Ultimate strength.

   (e) Breaking strength.

   (f) Percent elongation (do this by the last strain reading as well as by measuring the burnished marks).

   (g) Toughness (area under the stress–strain curve).

   (h) Specific strength (Utilize references for determining density).

(i) Specific stiffness.

Generate a table detailing all tensile test results. If a value is left out, note why.

3. Determine the uncertainty for stress and the uncertainty for strain for the unknown sample.

4. Identify the unknown material tested by referring to handbook, textbook, online, or archival data. Is there enough resolution in the test procedure to definitively identify the unknown metal, or are more than one possible types of material indicated by the results? Use the uncertainty developed in 3 to support your conclusions.

5. Describe the failure surfaces for each specimen. Take pictures of the failure surface. Discuss the failure mode for each of the tested materials.

6. When should you use the extensometer for measuring strain rather than the cross head displacement? What advantages does the extensometer have over cross head extension?

7. Which material has the highest specific strength and which has the highest specific stiffness? Note maximums in the table you have prepared.

3.5.7   PRINCIPAL EQUIPMENT REQUIREMENTS AND SOURCING

- UTS Machine with extensiometer.

- Various dogbone tensile specimens, Typical Vendor: Laboratory Devices CO, Inc.

- Uniaxial Carbon fiber laminate 0.35 mm thick: Typical Vendor
  http://www.cstsales.com/.

- Plaster of Paris cylinder, ~2.5 cm diameter, 4–6 cm long.

- Measurement tools in lab (rulers, micrometers, etc.). Various.

LABORATORY   4

Thin-walled Pressure Vessels

PART A: THEORY

4.1   THIN-WALLED PRESSURE VESSEL AND STRAIN ROSETTE

4.1.1   INTRODUCTION

Pressure vessels are used in industry to store fluids under high pressure. The vessels often have cylindrical or spherical geometry. Two examples of cylindrical pressure vessels to store gas are shown in Fig. 4.1. The wall thickness of the vessel is assumed to be small compared to its diameter and hence we use thin-walled pressure vessel theory to analyze stress in the material. Examples of thin-walled pressure vessels include pipes carrying fluids, scuba tanks, fire extinguishers, beverage cans, gas tanks, boilers, etc. The internal pressure causes stress in the pressure vessel wall and when this stress exceeds the strength of the material it will rupture catastrophically. So, determining the internal pressure within a pressure vessel is essential for its safe use.

Figure 4.1: Pressure vessels for storage of gas.

In this laboratory, we will utilize a strain rosette to determine internal pressure in a pressurized beverage can. A strain rosette consists of two or more strain gages on a single tab (backing material). The student will bond a 0°–45°–90° strain rosette on the surface of a can at *a random angle* to its longitudinal axis, and utilize strain gage theory coupled with pressure vessel theory to determine the internal pressure as well as the stress induced in the wall. Since the measurements being made are "strains" in random orientations and the quantity to be determined is "pressure" (or stress), we need to understand the relationship between strain and stress (pressure). Immediately, a student may think of Hooke's law, i.e., $\sigma_x = E\varepsilon_x$. Recall that this relationship is applicable for 1-dimensional stress state only, i.e., when the applied load acts along one direction on a well-defined specimen geometry, such as a dog-bone specimen you have used in a previous laboratory exercise when uniaxial stress–strain response was obtained. However, in the case of pressure vessels, the internal pressure acts in all directions on the container wall and so the above stress–strain relationship is inadequate to describe the complex stress state in the wall. Pressure vessels experience multi-axial stress. In general, thick-walled pressure vessels develop 3D stress state. However, in this class we will limit to **thin-walled** pressure vessels ($t \ll D$), such as beverage cans. Due to small thickness of the wall, it is assumed that there is no stress in thickness direction (i.e., 2D stress state) and the material experiences only in-plane stresses, i.e., **plane stress condition**, where σ_x, σ_y, and τ_{xy} may exist at any point on the surface of the can and all other stress components are zero (i.e., $\sigma_z = \tau_{yz} = \tau_{xz} = 0$). Thus, to analyze stress state in the pressure vessel wall and determine the pressure in the soda can, we use theory of *strain transformations* (convert measured strains in random orientation to principal strains) or use Mohr's circle, *2D stress–strain relationships* (convert strains to stresses), and *thin-walled pressure vessel theory* (relate calculated stresses developed in the wall to internal pressure).

4.2   THEORY OF STRAIN ROSETTE

In Fig. 4.2, a strain rosette is oriented at an arbitrary angle α to the horizontal (H) and vertical (V) axes. Here we use a $0°–45°–90°$ strain rosette to measure strains on a beverage can. Note that the outer two strain gages are at $0°$ and $90°$ and hence they form a convenient rectangular coordinate system with axes denoted as **x** and **y**. The x-axis is oriented at an angle α to the H-axis. The student will bond a $0°–45°–90°$ strain rosette at an arbitrary angle α to the axes of the cylindrical pressure vessel. Recall that each strain gage measures strain along its length direction. For a $0° - \theta° - 90°$ strain rosette, the following equations are relevant. Upon application of pressure, the following strains are measured by the strain gages along these directions:

$$\varepsilon_x = \varepsilon_{\theta=0} = \varepsilon_0 \tag{4.1}$$

$$\varepsilon_\theta = \varepsilon_x \cos^2\theta + \varepsilon_y \sin^2\theta + \gamma_{xy} \sin\theta \cos\theta$$

$$= \frac{\varepsilon_x + \varepsilon_y}{2} + \frac{\varepsilon_x - \varepsilon_y}{2}\cos 2\theta + \frac{\gamma_{xy}}{2}\sin 2\theta \tag{4.2}$$

$$\varepsilon_y = \varepsilon_{\theta=90} = \varepsilon_{90} \tag{4.3}$$

Although we have noted that the strain rosette is bonded at an arbitrary angle to the axis of the cylindrical beverage container, for convenience, we have chosen the x-axis as the axis of strain gage-1 ($\theta = 0°$) irrespective of its orientation to the axis of the container. Also, for the middle gage, the generalized equation for strain in any orientation θ is given in Eq. (4.2), although in the strain rosette chosen here, the middle strain gage is at $\theta = 45°$. In general, shear strain (γ_{xy}) cannot be measured directly by a single strain gage because each strain gage can measure only axial strain in the direction of gage wire lay up and hence it needs to be calculated from Eq. (4.2).

In this laboratory, we use strain measurements from the three gages in the strain rosette and Eqs. (4.1)–(4.3) to calculate shear strain (γ_{xy}). Both ε_x and ε_y are measured directly during the experiment. Knowing the value of strain from strain gage at $\theta = 45°$, i.e., $\varepsilon_{\theta=45}$, and

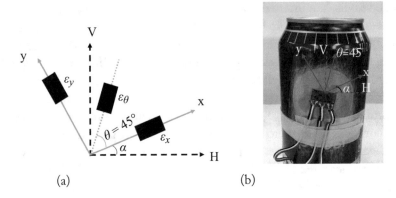

Figure 4.2: (a) Strain rosette with gages aligned at an arbitrary angle to H and V coordinate axes and (b) picture of a beverage can with strain rosette installed.

plugging in the values from $\varepsilon_x, \varepsilon_y$ and at strain at $\theta = 45°$ into Eq. (4.2), γ_{xy} can be calculated. Thus, we have complete state of strain at a point on the surface of the pressure vessel as indicated in Fig. 4.2. γ_{xy} is not indicated as it represents the change in angle between two originally orthogonal lines.

However, the strains ε_x and ε_y measured at a point on the surface of the vessel (Fig. 4.2) are oriented at an angle α to the H and V axes, as indicated in Fig. 4.3. We are often interested in principal strains because they provide the maximum (ε_1) and minimum (ε_2) strains. We can now calculate principal strains at that point from the above-measured strains using the following equations:

$$\varepsilon_{1,2} = \frac{\varepsilon_x + \varepsilon_y}{2} \pm \sqrt{\left(\frac{\varepsilon_x - \varepsilon_y}{2}\right)^2 + \left(\frac{\gamma_{xy}}{2}\right)^2} = \varepsilon_{avg} \pm \frac{\gamma_p}{2} \tag{4.4}$$

i.e.,

$$\varepsilon_1 = \varepsilon_{avg} + \frac{\gamma_p}{2} \quad \text{and} \quad \varepsilon_2 = \varepsilon_{avg} - \frac{\gamma_p}{2} \tag{4.5}$$

Also,

$$\frac{\varepsilon_1 + \varepsilon_2}{2} = \varepsilon_{avg} = \frac{\varepsilon_x + \varepsilon_y}{2} \tag{4.6}$$

Here, γ_p is the in-plane maximum shear strain given by

$$\gamma_p = \gamma_{\text{in-plane max}} = 2\sqrt{\left(\frac{\varepsilon_x - \varepsilon_y}{2}\right)^2 + \left(\frac{\gamma_{xy}}{2}\right)^2} = \pm(\varepsilon_1 - \varepsilon_2) \tag{4.7}$$

Note that this in-plane maximum shear strain is at 45° to the principal strain axes.

As will be seen below in the theory on thin-walled pressure vessel that these principal strains are oriented at an angle α to the x- and y-axes in a cylindrical pressure vessel, and this

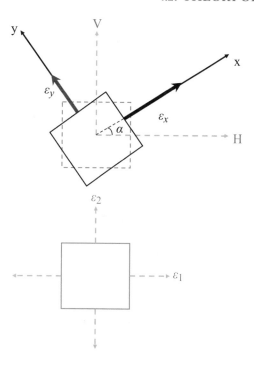

Figure 4.3: Orientation of measured strains ε_x and ε_y, and principal strains ε_1 and ε_2 (along H and V) at a point on the surface of a cylindrical pressure vessel.

angle is given by,

$$\alpha = \frac{1}{2}\tan^{-1}\frac{\gamma_{xy}}{\varepsilon_x - \varepsilon_y} \tag{4.8}$$

For plans-stress condition (i.e., $\sigma_z = 0$), the strain in the z-direction (out-of-plane) is given by

$$\varepsilon_z = \varepsilon_3 = -\frac{\nu}{1-\nu}\left(\varepsilon_x + \varepsilon_y\right) = -\frac{\nu}{1-\nu}(\varepsilon_1 + \varepsilon_2) \tag{4.9}$$

In Eq. (4.9) we have used the expression $\left(\varepsilon_x + \varepsilon_y\right) = (\varepsilon_1 + \varepsilon_2)$ from Eq. (4.6). Thus, we have now determined all the strains at this point and their orientations. But we need principal stresses to determine the pressure and maximum stress in the wall. So, we need to convert strains to stresses using 2D (biaxial, i.e., $\sigma_z = 0$) Hooke's law.

4.3   STRESS–STRAIN RELATIONSHIPS

In 2D, the stress–strain equations are given by

$$\varepsilon_x = \frac{\sigma_x}{E} - \frac{v}{E}(\sigma_y + \sigma_z) \xrightarrow{\sigma_z=0} \varepsilon_x = \frac{1}{E}(\sigma_x - v\sigma_y)$$

$$\varepsilon_y = \frac{\sigma_y}{E} - \frac{v}{E}(\sigma_x + \sigma_z) \xrightarrow{\sigma_z=0} \varepsilon_y = \frac{1}{E}(\sigma_y - v\sigma_x) \qquad (4.10)$$

$$\varepsilon_z = \frac{\sigma_z}{E} - \frac{v}{E}(\sigma_x + \sigma_y) \xrightarrow{\sigma_z=0} \varepsilon_z = -\frac{v}{E}(\sigma_x + \sigma_y)$$

From Eqs. (4.9) and (4.10), we can write,

$$\varepsilon_z = -\frac{v}{1-v}(\varepsilon_x + \varepsilon_y) = -\frac{v}{E}(\sigma_x + \sigma_y)$$

Also, note that shear strain is converted to shear stress as per

$$\gamma_{xy} = \frac{\tau_{xy}}{G}$$

where, G is shear modulus, E is the Young's modulus, and v is the Poisson's ratio.
    Inverting the above equations, we get stresses in terms of strains, i.e.,

$$\sigma_x = \frac{E}{1-v^2}(\varepsilon_x + v\varepsilon_y). \quad \text{Similarly,} \quad \sigma_1 = \frac{E}{1-v^2}(\varepsilon_1 + v\varepsilon_2)$$

$$\sigma_y = \frac{E}{1-v^2}(\varepsilon_y + v\varepsilon_x). \quad \text{Similarly,} \quad \sigma_2 = \frac{E}{1-v^2}(\varepsilon_2 + v\varepsilon_1) \qquad (4.11)$$

$$\tau_{xy} = G\gamma_{xy}$$

Also, note from Eqs. (4.1)–(4.6) we can write similar relations for stress by simply replacing strain with corresponding stress as shown below:

$$\varepsilon_x \rightarrow \sigma_x, \quad \varepsilon_1 \rightarrow \sigma_1$$

$$\varepsilon_y \rightarrow \sigma_y, \quad \varepsilon_2 \rightarrow \sigma_2 \qquad (4.12)$$

$$\frac{\gamma_{xy}}{2} \rightarrow \tau_{xy}, \quad \frac{\gamma_{max}}{2} \rightarrow \tau_{max}$$

If the strain rosette is randomly oriented at angle α to the horizontal or vertical axis of the pressure vessel, then the principal stress direction from x-axis can be determined by using

$$\theta_p = \frac{1}{2}\tan^{-1}\frac{2\tau_{xy}}{\sigma_x - \sigma_y} \qquad (4.13)$$

As will be seen in Section 4.4, that the principal stresses on a cylindrical pressure vessel are aligned along H and V axis. Thus, $\alpha = \theta_p$. The stresses and their orientations at any point

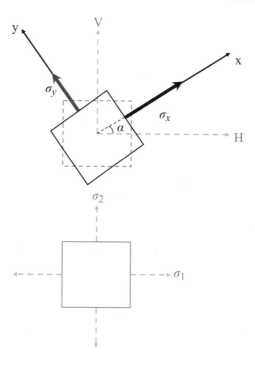

Figure 4.4: Orientation of principal stresses with respect to the strain rosette.

on the surface of the pressure vessel are shown in Fig. 4.4. Also, in Eq. (4.11), we have shown equations for principal stresses σ_1 and σ_2. What about σ_3? Because of plane stress condition on the surface of the can, $\sigma_3 = 0$.

Thus, we have only narrated how the measured strains from a strain rosette can be converted to principal stresses. We still need to relate these stresses to those developed in a pressure vessel.

4.4 THEORY OF THIN-WALLED PRESSURE VESSEL

Thin-walled pressure vessel, such as a beverage can or a gas tank (scuba cylinder), holds fluids under internal pressure. The pressure vessel wall develops a biaxial state of stress. So, we have to use biaxial stress–strain equations as those given in Section 4.3. In a cylindrical pressure vessel, two stresses are developed: axial stress or longitudinal stress (in the direction of cylindrical axis) and hoop stress (along the circumference). Thus, the stress state is biaxial. Because we measure stresses on the surface, we use plane stress condition, i.e., $\sigma_z = 0$. A special case of plane stress is the uniaxial stress tests you have performed in a previous lab, where a tensile specimen was pulled in one direction. In this tensile test, we have $\sigma_x > 0$, $\sigma_y = 0$, and $\sigma_z = 0$. In biaxial condition, such as the one on a pressure vessel surface, we have non-zero values for both σ_x and σ_y, but $\sigma_z = 0$. Student should recall that in both uniaxial stress and biaxial stress situations, all three strain components $(\varepsilon_x, \varepsilon_y, \varepsilon_z)$ exist.

Due to the applied internal pressure, cylindrical pressure vessels often burst (fail) by developing a crack along their length (longitudinal) direction, as shown in Fig. 4.5. To analyze

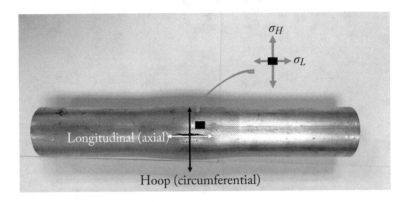

Figure 4.5: Failure of a pipe internally pressurized with fluid. Note that the crack occurs along the longitudinal (axial) direction because hoop stress (σ_H) is greater than longitudinal stress (σ_L).

the stresses developed in these vessels and determine the orientation of the maximum stress, we need to relate the internal pressure (P) to stress generated in the wall. Because we have noticed fracture along the longitudinal direction, let us analyze the free body diagram (FBD) shown in Fig. 4.6 for a cylinder of length L, wall thickness t, and radius r. Applying equations of equilibrium for forces perpendicular to the failure surface in the figure, we can write

$$(2rL).P = (2Lt).\sigma_H \tag{4.14}$$

Here, the left-hand side is the force exerted by the pressure on the cross sectional projected area $(2r.L)$ of the entire pressure vessel and the right-hand side is the resistance offered by the vessel wall whose area is $(2L.t)$. This resistance produces the hoop stress, σ_H, which from Eq. (4.14), is written as

$$\sigma_H = \frac{Pr}{t} \tag{4.15}$$

Similarly, we can draw a FBD for axial stress if we imagine failure to occur along the circumference of the pressure vessel. Applying equations of equilibrium for forces along the axial (longitudinal) direction, as shown in Fig. 4.7, we can write

$$P.(\pi r^2) = \sigma_L.(2\pi rt) \tag{4.16}$$

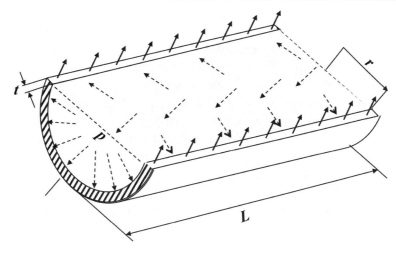

Figure 4.6: Forces acting on the failure plane of a cylinder cut along the longitudinal plane.

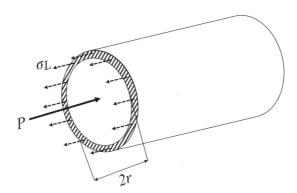

Figure 4.7: Forces acting on the failure plane of a cylinder cut along the circumferential plane.

Therefore,

$$\sigma_L = \frac{Pr}{2t}$$

(4.17)

Thus, in a cylindrical pressure vessel, axial and hoop stresses are developed which are given by Eqs. (4.15) and (4.17). If we can measure one of these stresses, the internal pressure P in the cylinder can be calculated.

By comparing Eqs. (4.15) and (4.17), we note that

$$\sigma_L = \frac{\sigma_H}{2} \quad \text{or} \quad \sigma_H = 2\sigma_L$$

(4.18)

i.e., hoop stress is twice the magnitude of the longitudinal stress. This is an important result which helps to understand the failure we have observed in Fig. 4.5. When internal pressure is applied, both stresses develop, but the magnitude of hoop stress is twice that of longitudinal stress. As the cylinder bulges, the hoop stress in the circumferential direction exceeds the strength of the material and hence a crack nucleates on a plane perpendicular to this stress direction and the cylinder bursts with crack growth along a plane parallel to the cylinder axis.

To better understand the stress state on the surface of the pressure vessel, we now represent these stresses using a Mohr's circle. Recall that σ_H and σ_L are along the hoop (circumferential) and axial (longitudinal) directions which are perpendicular to each other. These are the stresses on horizontal and vertical planes at any point on the surface of the pressure vessel. There is no shear stress on these two planes (shear stress can still exist on any other plane in different orientation). Therefore, by definition, they are the principal stresses (σ_1 and σ_2) at any location on the surface of the vessel, i.e.,

$$\sigma_H = \sigma_1 \quad \text{and} \quad \sigma_L = \sigma_2 \tag{4.19}$$

from Eq. (4.18), $\sigma_H = 2\sigma_L$, and therefore, $\sigma_1 = 2\sigma_2$.

Note from Fig. 4.2 that the two outer gages in the strain rosette are perpendicular to each other (along x- and y-axes) and are oriented at an angle α to the horizontal axis. The principal stresses are along horizontal and vertical axes of the pressure vessel (see Fig. 4.8a). Now, one

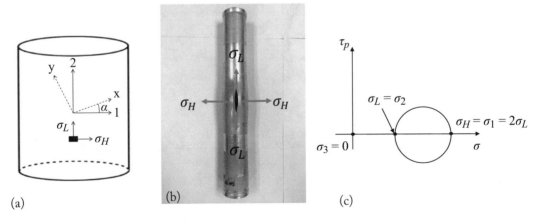

(a)   (b)   (c)

Figure 4.8: (a) Illustration of stresses acting on a cylindrical pressure vessel, (b) failure plane with respect to hoop stress, and (c) Mohr's circle representation of stress state at any point on the surface.

can calculate the angle between the x-axis of the strain rosette and the hoop stress (horizontal direction) of the vessel by using the Eq. (4.13).

$$\alpha = \theta_p = \frac{1}{2}\tan^{-1}\frac{2\tau_{xy}}{\sigma_x - \sigma_y} \tag{4.20}$$

where, σ_x, σ_y, and τ_{xy} are the stresses calculated using stress–strain relations [Eq. (4.11)] and the measured strains in Eqs. (4.1)–(4.3).

Figure 4.8 also illustrates that if the strain rosette was aligned along axial and hoop directions, the 45° strain gage is redundant because the x- and y-strain gages directly provide the principal strains and one can use Eq. (4.11) to obtain the principal stresses, i.e., σ_L and σ_H. Equations (4.13) or (4.20) can also be used to verify if the rosette is well aligned with H-axis (hoop) and V-axis (longitudinal) of the pressure vessel by calculating the principal stresses from the measured strains and checking the relationship $\sigma_H = 2\sigma_L$. Otherwise, the rosette is slightly off angle.

To determine if the pressure vessel fails due to applied pressure, calculate the hoop stress for given pressure and dimensions of the vessel using Eq. (4.15), and check if the calculated hoop stress exceeds the strength of the can material. These two stresses can now be represented on a Mohr's circle, as shown in Fig. 4.8c.

So far, we have focused on 2D stress state, i.e., under plane stress condition we only have two in-plane stresses, σ_x and σ_y (and $\sigma_z = 0$) or σ_1 and σ_2 (and $\sigma_3 = 0$). The in-plane maximum shear stress is given by

$$\tau_p = \tau_{\text{in-plane max}} = \pm\frac{\sigma_1 - \sigma_2}{2} = \pm\frac{\sigma_H - \sigma_L}{2} \tag{4.21}$$

were τ_p is the radius of the Mohr's circle with diameter $(\sigma_1 - \sigma_2)$. Because we have plane stress condition, this plane of maximum shear stress acts on a plane at 45° to the principal planes which are horizontal (circumferential plane) and vertical (longitudinal or axial) planes, as shown in Fig. 4.9a in a 2D diagram and in Fig. 4.9b in a 3D diagram. The relevant Mohr's circle is shown once again in Fig. 4.9c.

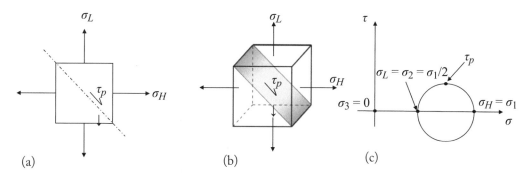

Figure 4.9: (a) The stress state at any point on a cylindrical pressure vessel wall with its hoop and longitudinal stresses indicated in 2D, (b) the plane of maximum shear stress in 3D representation, and (c) illustration of stresses on Mohr's circle at the same point.

However, the maximum shear stress at that point is not τ_p because at that point σ_3 also acts whose magnitude is zero (by definition of plane stress condition) and this stress acts perpendicular to the surface; see Fig. 4.10b. Because maximum shear stress is defined as half the difference between the maximum and minimum principal stresses acting at that point, it is given by

$$\tau_{max} = \frac{\sigma_1 - \sigma_3}{2} = \frac{\sigma_1 - 0}{2} = \frac{\sigma_H}{2} > \tau_p = \frac{\sigma_1 - \sigma_2}{2} = \frac{\sigma_H - \sigma_L}{2} = \frac{\sigma_H - \sigma_H/2}{2} = \frac{\sigma_H}{4}$$
(4.22)

Thus,

$$\tau_{max} = \frac{\sigma_H}{2} \quad \text{whereas,} \quad \tau_p = \frac{\sigma_H}{4}, \quad \text{i.e.,} \quad \tau_{max} = 2\tau_p$$
(4.23)

These relations can be visualized from the Mohr's circles in Figs. 4.9c and 4.10c. So, where does this plane of maximum shear stress act? Once again, we follow the approach that we adopted for τ_p. Because τ_{max} is derived from, σ_1 and σ_3, we note from Mohr's circle that this plane of max shear is at 45° to the planes on which σ_1 and σ_3 act, as depicted in Fig. 4.10.

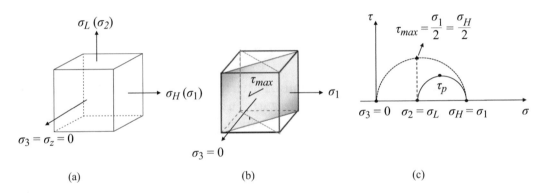

(a)   (b)   (c)

Figure 4.10: (a) 3D stress state at any point on a cylindrical pressure vessel wall, (b) the plane of maximum shear stress in 3D representation, and (c) illustration of in-plane maximum shear stress (smaller circle) and out-of-plane maximum shear stress on two Mohr's circles at the same point.

Comparing the two Mohr's circles in Fig. 4.10c, it is seen that the radius of the circle that represents in-plane stresses is half the size of the radius of circle that represents the out-of-pane stresses. Clearly, the plane of in-plane max shear stress (Fig. 4.9b) is different from that of the plane of maximum shear stress (Fig. 4.10b) at that point. *So where does failure occur in a pressure vessel when it is pressurized?* Obviously, it occurs wherever the max shear stress exists. From the above figure, it is the plane of maximum shear stress (τ_{max}) which is at 45° to the surface of the pressure vessel but going in the thickness direction, precisely the orientation of the fracture plane noticed on the pipe shown in Fig. 4.11.

Figure 4.12 shows another example of a scuba tank that exploded due to inappropriate heat treatment which caused the strength of the material to degrade dramatically. When pressurized

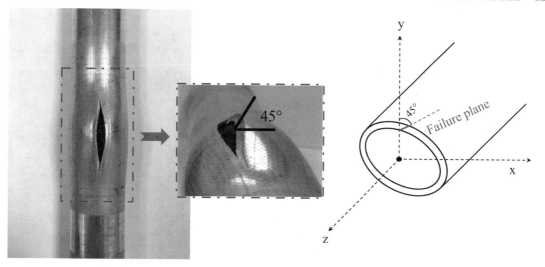

Figure 4.11: Orientation of the fracture plane in thin wall pressure.

to its original rated capacity, it exploded. Once again note the fracture along the axial direction (due to hoop stress). Interestingly, the failure plane is at 45° to the thickness, similar to Fig. 4.11 and also as predicted by the theory in Fig. 4.10. Another interesting observation in the failure plane is that there is a transition of failure mode from plane stress condition to plane strain condition (can be seen in the image on the right). Recall that our theory in this laboratory pertains to thin walled pressure vessel where the thickness of the wall is assumed to be small and hence the stress-state is assumed not to change over thickness. The stress in the thickness direction is assumed to be small and constant (plane stress condition). However, the scuba tank thickness is much greater and hence the thin-walled pressure vessel theory is not fully applicable.

Now that we have understood the failure theory as applicable to pressure vessels, let us do some brainstorming.

Questions:

1. Can we use a 0°–90° strain rosette to measure pressure in the can?

   The answer is yes. From the measured strains in *any* two perpendicular directions (say, x' and y'), we can calculate normal stresses ($\sigma_{x'}$ and $\sigma_{y'}$) along those directions using Eqs. (4.11). From Mohr's Circle theory, we know that, the center of the Mohr's circle is at a distance from the origin given by:

$$\frac{\sigma_{x'} + \sigma_{y'}}{2} = \frac{\sigma_x + \sigma_y}{2} = \frac{\sigma_1 + \sigma_2}{2} = \frac{\sigma_H + \sigma_L}{2} = \frac{\sigma_H + \sigma_{H/2}}{2} = \frac{\frac{3}{2}\sigma_H}{2}$$

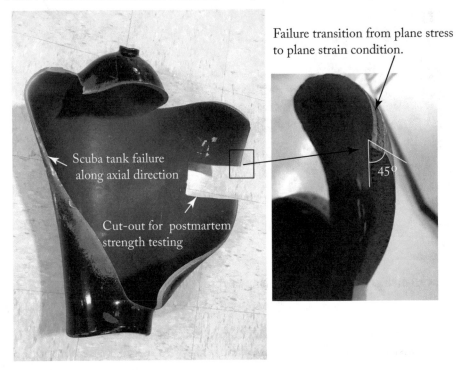

Figure 4.12: Exploded scuba tank failure along axial direction. A small cutout on the right was made to test the strength of the failed material. The image on the right shows the fracture angle which is at 45° to the thickness direction. The failure plane transition from plane stress to plane strain can also be noted.

therefore,

$$(\sigma_{x'} + \sigma_{y'}) = \frac{3}{2}\sigma_H = \frac{3}{2}\left(\frac{Pr}{t}\right)$$

thus,

$$P = (\sigma_{x'} + \sigma_{y'})\frac{2t}{3r} \tag{4.24}$$

Therefore, regardless of the orientation, one can use a 0°–90° strain rosette to calculate the pressure in the can.

2. Now consider a situation where only one gage is available. Can we determine the pressure in the can?

Again, the answer is yes because we know that the magnitudes of the hoop and longitudinal stresses are not independent but they are related, i.e., $\sigma_H = 2\sigma_L$. Therefore, if we know one of the stresses, the other can be determined. Also, these two stresses are the principal stresses and, hence, they are perpendicular to each other. Therefore, by bonding the strain gage along one of the principal axes and using the biaxial stress–strain relations in Eq. (4.11), we can write,

$$\varepsilon_1 = \frac{1}{E}(\sigma_1 - v\sigma_2) = \frac{1}{E}(\sigma_H - v\sigma_L) = \frac{1}{E}\left(\frac{Pr}{t} - v\frac{Pr}{2t}\right) = \frac{Pr}{Et}\left(1 - \frac{v}{2}\right)$$

$$\varepsilon_1 = \frac{Pr}{2Et}(2 - v)$$

which gives,

$$\boxed{P = \frac{2Et\varepsilon_1}{r(2 - v)}} \tag{4.25}$$

Thus, knowing dimensions of the pressure vessel (t and r) and its properties (E and v), we can calculate the internal pressure by just measuring one principal strain.

3. In Fig. 4.2, the strain rosette has gages oriented along 0°–45°–90° which makes the Eqs. (4.1)–(4.3) simpler. What changes need to be made if the three gages were at an arbitrary angle to each other? Can you determine the pressure in the gage? Explain.

4.5   UNCERTAINTY CALCULATIONS (FROM HOOP STRESS)

In this lab the pressure in a beverage can is determined by measuring strains on the surface of a pressure vessel, i.e., measure, ε_x, ε_y, and ε_θ, calculate stresses, σ_x, σ_y, and τ_{xy} and then principal stresses, σ_1, σ_2, and τ_p. If ε_x and ε_y are along horizontal and longitudinal directions then

$$\begin{cases} \sigma_x = \sigma_1 = \sigma_H \\ \sigma_y = \sigma_2 = \sigma_L \end{cases}$$

Finally, the pressure is calculated from either hoop or longitudinal stresses, i.e.,

$$\sigma_H = \frac{Pr}{t} \quad \text{which yields} \quad P = \frac{\sigma_H t}{r}$$

The uncertainty in the pressure is given by

$$U_P = \left[\left(\frac{\partial P}{\partial \sigma_H}\right)^2 U_{\sigma_H}^2 + \left(\frac{\partial P}{\partial t}\right)^2 U_t^2 + \left(\frac{\partial P}{\partial r}\right)^2 U_r^2\right]^{\frac{1}{2}} \tag{4.26}$$

where uncertainty in hoop stress U_{σ_H} can be calculated from

$$\sigma_H = \frac{E}{1 - v^2}(\varepsilon_1 + v\varepsilon_2) \quad \text{where} \quad v = 0.3$$

$$U_{\sigma_H} = \left[\left(\frac{\partial \sigma_H}{\partial E} \right)^2 U_E^2 + \left(\frac{\partial \sigma_H}{\partial \varepsilon_1} \right)^2 U_{\varepsilon 1}^2 + \left(\frac{\partial \sigma_H}{\partial \varepsilon_2} \right)^2 U_{\varepsilon 2}^2 \right]^{\frac{1}{2}} \tag{4.27}$$

If the strain gage is aligned along the principal axis, the principal strain is given as

$$\varepsilon_1 = \frac{4(\Delta V)}{V_s G_f} \tag{4.28}$$

Now, calculate the uncertainty in measured strain and complete the steps.

Sample Problems

1. A student bonds a 0°–45°–90° strain rosette on a pressure vessel at random angle with an intent to measure pressure in the can. While connecting the strain gages to lead wires, he damaged the middle gage which is at 45°. Knowing the theory for pressure vessels very well, the student continued with the experiment and measured the strains from the two outer gages when the vessel was pressurized. The measured strains are 720×10^{-6} and 330×10^{-6}. The pressure vessel was made of steel (Young's modulus = 210 GPa and Poisson's ratio = 0.3) and has a diameter of 500 mm and wall thickness of 3 mm.

   (a) Calculate the pressure in the can?

   (b) Calculate the Hoop and Longitudinal stresses in the pressure vessel wall?

   (c) What are the principal stresses at the location of the strain gage?

   (d) What is the absolute maximum shear stress at the location of strain gage?

   (e) On a Mohr's circle indicate all the above stresses.

   (f) What is the stress along the thickness of the can?

   (g) What is the strain along the thickness direction of the can?

2. A student uses a single strain gage (gage factor = 2.0) to measure the pressure in a cylindrical beverage can (dia = 100 mm and thickness = 0.25 mm) made of aluminum (Young's modulus = 70 GPa and Poisson's ratio = 0.25). The strain gage is bonded horizontally on the circumferential surface and connected to a W-bridge in 1/4-bridge arrangement. A 10 V DC supply is given to the bridge. The can is now opened and the output voltage of 0.5 mv is measured. No amplifier is used. Determine the pressure in the can.

PART B: EXPERIMENT

4.6   STRAIN ROSETTE BONDING AND DETERMINATION OF PRESSURE IN A BEVERAGE CAN

4.6.1   OBJECTIVE

Determination of pressure in a beverage can (pressure vessel) using a strain rosette. In this lab, the student will bond a 0°–45°–90° strain rosette to a beverage can and determine the stress on the surface of the can. From the calculated stresses, the internal pressure in the can should be determined.

Prelab Question

Find the strains a 0°–45°–90° strain rosette will measure when placed at an arbitrary angle with respect to the cylindrical axis (not 0° or 90°) on a beverage can with an internal pressure of 200 kPa. Report the angle and the three strains. If more data is needed, list your approximations and reference.

Week 1:

4.6.2   EQUIPMENT AND SUPPLIES NEEDED

- Strain rosette.

- Soda can.

- Strain gage mounting supplies (see Fig. 4.13).

4.6.3   EXPERIMENTAL TASK

1. Bond the strain rosette to the beverage can. Ensure that the strain rosette is at a random angle to the horizontal axis, as shown in Fig. 4.2b.

   (a) Adapt the instructions used for strain gage installation in Lab 2 (Vishay Strain Gage Installation Manual (Instruction Bulletin B-27-14, `http://www.vishaypg.com`).

   (b) Use 3, 3-wire hook-ups (we are using 3 single-gage strain gage amplifiers).

   (c) Ensure the resistances are within nominal values.

   (d) Place names on tape on can; set aside for use next week.

Week 2:

Before attending the lab, develop a VI that will estimate and save the strains acting on the can over time. Include the necessary steps to evaluate uncertainties in the process.

Figure 4.13: Beverage can, strain rosette, and supplies needed for the experiment.

4.6.4 EQUIPMENT NEEDED

- Laptop computer with LabVIEW installed.

- Multi-function DAQ.

- 3 1/4-bridge strain gage amplifiers.

- Strain rosette mounted on a soda can.

- Snips for cutting can.

- Micrometer for measuring can thickness.

- Multi-meter.

4.6.5 REQUIRED LABVIEW PROGRAM (VI)

1. Write a LabVIEW program that uses the "Gain.vi" as a SubVI that will estimate and save the strains acting on the can over time. Include the necessary steps to evaluate uncertainties in the process.

(a) The DAQ has four differential analog inputs. Determination of the V_s voltages of each W-bridge will be done before the VI is run, either with a simple VI or a multimeter.

(b) Display and record strain and V_{amp}. For each of the three gages in the rosette:

    i. Consider the strain gage amplification factor.

    ii. The gage factor for most gages typically used is $2.1 \pm 0.5\%$.

4.6.6   EXPERIMENTAL TASK

1. Connect the strain gages to the amplifiers as done in Lab 2 (see Amplifier Documentation for details and Fig. 2.15). Three amplifiers are needed.

2. Establish the values of the excitation voltages of each amplifier (V_s), and input into the VI.

3. Run the VI and tare the initial readings.

    (a) Adjust the amplifiers so that the outputs will not saturate the analog to digital converters when the can is opened.

    (b) Tare the three ΔV_{amp} to zero for no disturbance of the can.

4. Run VI and gently squeeze the can. Take data and note observations. Use this step to de-bug the apparatus/VI.

5. Run VI and shake the can (carefully). Take data and note observations. Does the pressure increase?

6. Run VI and open the beverage can and measure the release of strain in the readings.

7. Empty can (drain or drink). Cut a sample from the can wall and measure the thickness with a micrometer.

8. Calculate the principal strains and maximum shear strain.

9. Calculate the principle stresses and maximum shear stress.

10. Which one of the two principle stresses is bigger (longitudinal or hoop stress?)

11. Plot all the above stresses and strains (both measured and calculated) on separate Mohr's circles.

12. Calculate and report the pressure estimated to an in lab TA to tabulate for further analysis by you (see discussion point #6 below). Include the type of beverage.

4.6.7   ISSUES TO BE DISCUSSED IN THE LAB REPORT

1. What is the effect of the orientation of the strain gage on measured pressure? (Does the orientation of strain gage matter and why?)

2. Verify if the relationship $\sigma_L = \sigma_H/2$ holds for your pressure vessel.

3. Use Mohr's circle to show that a two element 90° rosette at an arbitrary angle α is sufficient to determine the pressure in the can.

4. Explain how you can determine pressure in the can with a single strain gage and explain the rationale why this is possible.

5. Assume that you have attempted to bond the three-gage strain rosette with the first and third gages aligned with horizontal and vertical axes. How can you verify the exact orientation of the gages to be perfectly aligned with the two axes of the beverage can?

6. In your lab section, note how many types of beverage cans are being tested. Download the compiled pressure data (from #12 above) from all sections and develop 95% confidence intervals for the pressure in each type of beverage can. Discuss the results.

7. Discuss source of errors. Are the errors in measurements significant with respect to the confidence intervals developed in #5?

4.6.8   PRINCIPAL EQUIPMENT REQUIREMENTS AND SOURCING

- Multi-function DAQ with USB cable: Minimum 2 channel 14 bit ADC, USB interface, LabVIEW support. Typical vendor: Out of The Box SADI DAQ,

  `https://ootbrobotics.com/`.

- Student Strain Gage Rosette, 120 Ω, Typical Part: Micro-Measurements CAE-13-120CZ-120.

- Strain gage amplifier, 3 parts, typical part: Tacuna Systems EMBSGB200_2_3.

- Power supply: 6-12 VDC, source: various.

- Strain gage install materials, various.

- Multi-meter: Fluke 115 or similar.

- Measurement tools in lab (rulers, micrometers, etc.).

LABORATORY 5

Strength of Adhesive Joints

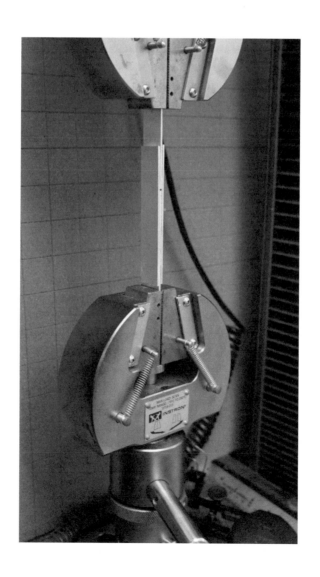

PART A: THEORY

5.1   SHEAR STRENGTH OF ADHESIVE JOINTS

5.1.1   INTRODUCTION

Adhesives have become increasingly popular as they require no mechanical fasteners (e.g., bolts, screws and rivets) to join two structural components. Adhesives are polymeric materials and have low density compared to metallic material they bond. In recent years, polymeric adhesives have been engineered to provide bond strength comparable to the strength of the parent components which they bond. Some adhesives are environmentally stable and are used extensively in automotive industry to reduce weight and improve fuel economy. Adhesives can also bond dissimilar materials. They are preferred candidates when the primary design requirement is high strength-to-weight ratio. Due to their polymeric nature, they have good damping properties and can be effective in reducing noise and vibration compared to traditional mechanical metal fasteners. Adhesive joints are also preferred because they do not cause stress concentration which can occur with mechanical fasteners such as rivets, bolts and nails.

Adhesive joints are most effective in transmitting shear and compression forces, and are relatively less effective in transmitting tensile forces. In this laboratory, we focus on determining shear strength of an adhesive when bonding two similar structural components. Adhesive strength of a joint made of metallic strips bonded by an adhesive is evaluated, either using single lap-shear (Fig. 5.1) or double lap-shear (Fig. 5.2) configuration. A load is applied on metal plates such that a state of pure shear is exerted in the adhesive. In both single- and double-lap shear tests, it is assumed that uniform shear force exists in the bonded region throughout the test process. In general, adhesive failure precedes the structural failure of the bonded components. However, when referring to the strength of an adhesive, we must distinguish between two kinds of strength: cohesive strength and adhesive strength. Adhesive strength refers to the strength of the bond between two dissimilar materials (adhesive and metal) whereas cohesive strength refers to strength between two similar materials (adhesive itself). For example, Fig. 5.1 illustrates three possible failure modes of an adhesive bond: adhesive failure, cohesive failure, and mixed mode failure. Adhesive failure occurs when the bond between the parent material and the adhesive is weak (Fig. 5.1a). Cohesive failure occurs the cohesive strength of the adhesive is lower than its bond strength with parent material (Fig. 5.1b). This could occur either because the adhesive thickness is large or because the adhesive may contain impurities or gas pockets etc. In Fig. 5.1a, the true bond strength of the adhesive is determined because the failure occurred between the adhesive and the parent material. Figure 5.1c is a mixed mode failure. This failure mode often occurs due to improper bonding or defects (air pockets or impurities) in the adhesive. It is important to recognize that the bond strength of an adhesive depends not only on the properties of the adhesive and the parent material being bonded, but also on surface properties such as cleanliness of surface, surface roughness, environmental conditions (temperature, humidity, time for curing of the bond), etc.

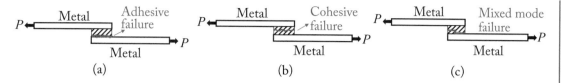

(a)   (b)   (c)

Figure 5.1: Schematic of single lap shear experiment of an adhesively bonded joint and illustration of failure modes: (a) adhesive failure, (b) cohesive failure, and (c) mixed mode failure.

Adhesive strength can also be determined using a double-lap shear test shown in the Fig. 5.2. Here three metal plates are bonded using the adhesive to be tested for its bond strength and the joint is subjected to shear forces as shown. The failure strength is determined by dividing the failure load by the area over which the adhesive has failed (debonded).

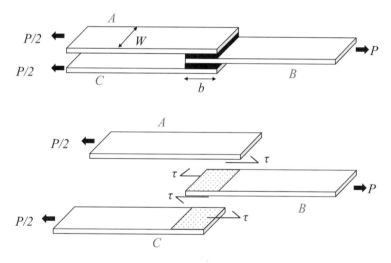

Figure 5.2: Schematic of the uniform shear stress experienced by an adhesive in a lap shear experiment.

By drawing free body diagrams as shown in Fig. 5.2, one can determine the area resisting the applied shear force to calculate the shear strength of the adhesive. The shear strength is given by

$$\tau = \frac{P/2}{A} = \frac{P}{2A} \tag{5.1}$$

where, P is the failure load and A is the area ($A = bW$) over which the adhesive has separated from the parent material. Note that regardless of the plate chosen in Fig. 5.2, the shear strength is given by Eq. (5.1). For upper and lower plates in Fig. 5.2, the force is $P/2$ and the shear area is A on each plate. For the middle plate, the force is P and the total area of the adhesive on both sides is $2A$. Thus the strength is given by $\tau = P/2A$.

However, this type of specimen is difficult to load due to asymmetry in the specimen geometry (loading on one plate on one side and two plates on the other side). Therefore, symmetric double-lap geometry is used such as the one shown in Fig. 5.3 which allows easy gripping of a single plate where the load P is applied.

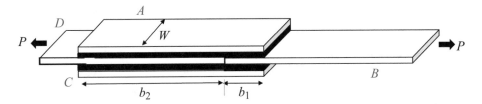

Figure 5.3: Schematic of symmetric adhesive bonded joint of metal plates.

Note that the central two plates are not bonded to the top plates exactly in the middle, but closer to one end of the upper plates. This arrangement will take away uncertainty in which side the bond will fail. Due to the smaller bonded area on the right side with area $W \times b_1$, the joint will fail on this side (assuming uniformity in bond strength all over the joint). The analysis for failure strength is similar to the one discussed earlier in relation to Fig. 5.2.

Sample Problem

A student intends to determine the shear strength of an adhesive that recently appeared in the market. He/she decides to use double lap shear configuration as shown in Fig. 5.3. All plates are of the same size with dimension of each plate being 2 mm × 20 mm × 100 mm. The location where the central plates meet edge-on is off set with respect to the center of the outer plates by 25 mm. The assembly is loaded until the joint broke at a force of 200 N. Assuming that one side of the joint is fully separated, determine the shear strength of the adhesive bond? Explain your analysis with a free body diagram.

PART B: EXPERIMENT

5.2   DOUBLE LAB SHEAR TESTING OF ADHESIVES

5.2.1   OBJECTIVES

In this laboratory, two commercially available adhesives are used to bond long strips of metallic materials as shown in Fig. 5.3. Three objectives are identified.

1. Perform standard double lap shear experiments on two samples of various adhesive chemistry.

2. Use optical techniques and image processing software to measure the adhesive contact area.

3. Quantify and contextualize confidence intervals for the shear strength of commercially available Cyanoacrylate (superglue) and Loctite Hysol E-05CL high-strength epoxy.

5.2.2   PRELAB QUESTION

Download and install ImageJ® software from `https://imagej.nih.gov/ij/download.htm l`. Your instructor will give three area images on which you will perform steps 5, 6, and 7 from the Experimental Tasks proposed in Section 5.2.4, and report the resulting areas. The class response will be provided to all to establish a range of uncertainty in the ImageJ area measurement process.

5.2.3   EQUIPMENT AND RESOURCES NEEDED

Instron 5567 and/or 5967 Universal testing Machine with 30 KN Load Cell, Eight strips of aluminum (four for each bonded joint) of 19 mm wide X 1.6 mm thick X 10 cm long to fabricate each shear specimens, Calipers and micrometers, cell phone camera, and ImageJ processing software on a computer. Cyanoacrylate (superglue) and Loctite Hysol E-05CL high-strength epoxy; see Fig. 5.4.

5.2.4   EXPERIMENTAL TASKS

Before your lab section, download and install ImageJ, `https://imagej.nih.gov/ij/download.html`.

1. Assemble two double-lap shear specimens, one with superglue and one with the Hysol epoxy. Take care in preparing the adhesive surfaces. Polish intended bond area of metal strips lightly with a fine sandpaper to get rid of any dust and clean with acetone or alcohol. Apply adhesive and press the two metal pieces to bond. Hold down for 3–5 min. Ensure the outside laps are symmetric with respect to each other. Ensure 2/3 of the length of the outside laps engage one interior aluminum strip, the balance the other; as shown in Fig. 5.5.

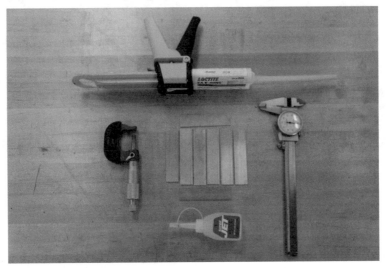

Figure 5.4: Equipment and supplies needed.

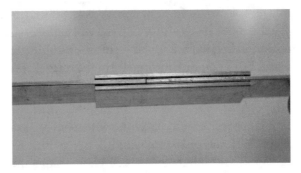

Figure 5.5: Bonded metals strips for testing of adhesive strength in double shear.

2. An instructor will configure UTS for tensile testing. Mount specimen in tensile grips in UTS (Fig. 5.6). Tare Machine. Set test speed at 7 mm/m. Run test until failure.

3. Observe double-lap shear tests on the two test specimens fabricated using two adhesives. Note failure mechanisms and observe the failure surfaces under a microscope.

4. Take pictures of the failure surfaces, with a scale included in the image.

5. Load the images into ImageJ.

6. Establish the scale (distance per pixel): Select the Choose the line tool and draw a line across a known length, on the scale marker (hold down the Shift key as you are drawing to align it horizontally or vertically). Analyze → Set Scale, ImageJ knows how many pixels

Figure 5.6: Double-lap shear test specimen loaded into UTS.

you selected with your line; type in the known distance and the units. Note the image calibration is listed on the bottom of that dialog box, and upon exiting that dialog box, image dimensions in the description are now in your units at the top of the image.

7. Draw an outline by selecting multiple points: Select the "Freehand Selections" tool and outline the area which best represents the contact area. Analyze → Measure to estimate the area of the bounded shape.

8. Note the dominating failure mechanism (adhesive vs. cohesive). Estimate the percent that this mechanism dominated by the ratio of areas of the two mechanisms, if possible.

9. As a class, input the maximum load each double-lap shear test sample experienced and the measured contact area of the failure into a spreadsheet. Include the observations from step 3 above. This data, along with all other sections' data, should be used for analysis by the student.

5.2.5   ISSUES TO BE DISCUSSED IN THE LAB REPORT

1. Explain why you have prepared the test specimen with 2/3 of the outside plates engaged with one interior aluminum strip (see Fig. 5.5).

2. Calculate the maximum shear stress that each double-lap shear specimen was able to withstand by dividing the maximum force by the area over which the adhesive was loaded before failure. Perform this for all specimens. Present this data in the Discussion section or an appendix.

3. Compute, tabulate, and report the 95% confidence interval for maximum shear stress that each specimen was able to withstand for each type of adhesive.

4. Discuss factors and their contribution to shear strength of the specimen. If you identify a trend, support your assertion using a plot or statistical description of data.

5. Develop and report an estimate of the uncertainty in the measurement of the failure area (Fig. 5.7) using ImageJ. Use the data provided from the prelab and report one standard deviation as the expected uncertainty.

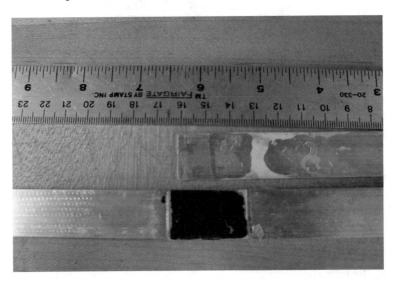

Figure 5.7: Failed bonded joint revealing the adhesive distribution on the joint. The glue area that failed has been inked to improve contrast (preform as necessary).

6. Discuss the order of magnitude of uncertainty in force and area used in calculating the shear stress. Compare and contrast with the statistical variation found in step 2. Are the uncertainties seen in here the result of inaccuracies in these measurements?

7. Find references for expected shear strength of CA and Hysol adhesives. Compare the results found with those you have determined in the lab and discuss.

Creep Behavior of Metals

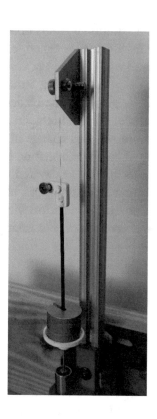

PART A: THEORY

6.1   INTRODUCTION

Mechanical behavior and strength of materials depend on many factors such as material initial microstructure, rate of applied load, temperature, time duration over which the a load is applied, etc. In this chapter, we will focus on deformation of a material under constant load applied over a long period of time. In Lab 3 (tensile stress–strain response of materials), we continuously applied increasing load to cause increasing elongation of a specimen. However, while obtaining "load-based stress–strain curve" using a thin copper wire, we were cautious not to keep weights for longer duration in order to prevent additional elongation in the wire due to constant load (or stress). This deformation under constant load is called "creep." Thus, "creep" is permanent deformation that occurs when constant load (or stress) or constant temperature is applied over an extended period. Creep occurs in many engineering structures. For example, telephone or power lines sag over time due to self-weight. Turbine blades in a jet engine undergo change in dimension when the engine operates at elevated temperature over an extended period. By definition, creep is an extremely slow process, and change in dimension of parts occurs over a long period. Therefore, understanding of creep deformation is important in many engineering applications especially when components operate at elevated temperatures such as in jet engine blades, boiler components, pipes and joints in oil refineries, etc. For efficient function, it is important that components operate over an extended period with minimal change in dimensions. Industry often sets a limit on allowable creep rates. For example, jet engines blades may have limits, for, e.g., 1% strain in 10,000 h.

6.2   MECHANISM OF CREEP

The dominant mechanisms for creep failure at atomic level is permanent movement of atomic planes at elevated temperature. Dislocation movement through atomic planes and grain rotation at high temperatures occur and eventually lead to failure of the material. Creep deformation can occur even in high strength and high heat-resistant materials. The strength of a material undergoing creep can be expressed as a function of strain rate (how fast or slow the load is applied) and time of exposure. These curves are called "creep curves," where strain is plotted as a function of time, as shown in Fig. 6.1. The slope of this curve is strain rate $\left(\frac{\Delta \varepsilon}{\Delta t}\right)$, which is also shown on this plot and has units of 1/s. The figure consists of three regimes: primary creep, secondary creep, and tertiary creep. Upon application of load, initial elastic deformation occurs followed by permanent plastic deformation at a slow rate (primary creep). Over time, strain increases slowly but steadily in a linear fashion (secondary creep) and eventually, the strain increases at a faster rate leading to failure. These three stages of creep are defined as follows.

Stage I – Primary Creep: Rapid increase in strain after initial elastic deformation. Material strain-hardens, and transitions to constant strain rate deformation (linear strain increase with time).

Stage II – Secondary Creep: This regime is also referred to as steady-state creep regime. Here the strain increases slowly but linearly with time. Hence, the slope (strain rate) is constant. In this regime, the material tries to recover while simultaneously strain hardening, i.e., there is a competition between strain hardening and strain recovery. Failure is not expected in this regime but strain continues to accumulate. From the design point of view, this is the preferred regime to operate a machine component because the secondary creep can occur over an extended period, and the linear nature of the curve with constant strain rate makes it easy to model mathematically and make prediction on the operational life of the component.

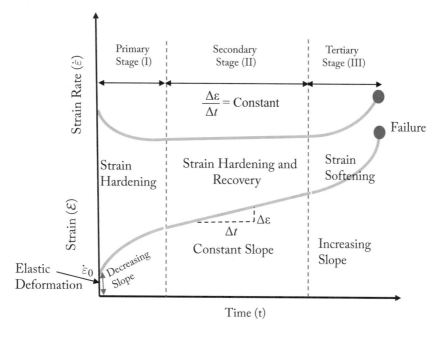

Figure 6.1: Typical creep curve showing the three regimes of creep deformation.

Stage III – Tertiary Creep: In this regime, the accumulated strain increases rapidly in an exponential fashion in a short duration. Simultaneously, strain rate also increases as the microstructural (or metallurgical) changes continue to accumulate in the material. These microstructural changes can be in the form of grain boundary separation, formation of cracks, cavities, voids, etc., leading to eventual failure of the material.

To develop creep curves, such as the one shown in Fig. 6.1 for any material, uniaxial tension tests, similar to the one you have performed in Lab 3, are conducted at a range of elevated temperatures. The difference is that in these tests a constant load is placed and the displacement is monitored over a long period. Thus, the test is conducted over several hours (low-strain rate). The form of the resulting curves are shown in Fig. 6.2. In these experiments, either temperature is increased at a constant stress level or stress is increased at a constant temperature. It is also possible to increase stress and temperature simultaneously. But these experiments are more complex. Depending on the test conditions, three possible scenarios exist. All equations correspond to steady state creep.

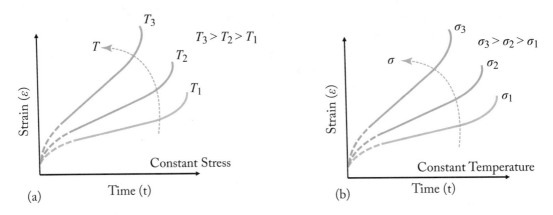

Figure 6.2: Creep strain curves under (a) constant stress and (b) constant temperature.

1. When temperature (T) is kept constant, and stress (σ) is varied, the creep rate in the steady state (secondary creep) regime can be defined as

$$\frac{d\varepsilon}{dt} = C\sigma^n \tag{6.1}$$

where, C and n are constants.

2. When temperature is varied and stress is kept constant, the steady-state creep rate is modeled as

$$\frac{d\varepsilon}{dt} = Ae^{\frac{-Q}{RT}} \tag{6.2}$$

where, A is constant, Q is activation energy, R is gas constant, and T is absolute temperature.

3. When both stress and temperature are varied, the steady-state creep rate is given by

$$\frac{d\varepsilon}{dt} = K\sigma^n e^{\frac{-Q}{RT}} \tag{6.3}$$

where, K and n are constant.

Typical creep curves for the first two scenarios are depicted in Fig. 6.2. Note that with increase in temperature or stress, the time to failure is reduced dramatically.

In your laboratory exercise, you will be asked to develop a creep curve at a constant load and at room temperature.

PART B: EXPERIMENT

6.3   CREEP BEHAVIOR OF A METALLIC WIRE

6.3.1   OBJECTIVE

This laboratory exercise is designed to examine the property of creep in a metallic wire. You will use known commercial weights to incrementally load a metal wire and measure elongation over time with a LVDT. Throughout the experiment, you are to monitor uncertainty and develop estimates of the accuracy and precision of your results.

6.3.2   PRELAB QUESTIONS

1. Calculate the weight at which the wire will yield if it is made of 30 AWG copper.

2. Select a weight that will exhibit creep and report for 30 AWG copper.

3. Using the weight from the step above, estimate an appropriate length of wire (30 AWG) so that the 20 mm LVDT travel is not an issue in the experiment, but resolution is appropriate. Discuss limitations this imposes on the experiment.

6.3.3   BACKGROUND NEEDED FOR CONDUCTING THE LAB

1. Review the MOM relationships associated with creep.

2. Familiarize yourself with LVDT's. See Lab 3 and:

   http://www.rdpe.com/us/hiw-lvdtdc.htm.

   We are using a RDP-sourced LVDT.

6.3.4   EQUIPMENT NEEDED

Same equipament as Lab 3: Load Controlled Stress–Strain Curve.

- Laptop computer with LabVIEW installed and functioning.

- Multi-function DAQ.

- Wire (copper).

- LDVT.

- Brass weights.

- Tensile loading fixture.

- Rulers, scales, dial calipers.

6.3.5   PROBLEM STATEMENT

Examine the creep properties of a wire. Use the information measured to develop characterization of creep for the material. Develop uncertainties for the procedure.

6.3.6   REQUIRED LABVIEW PROGRAM (VI)

Create a LabVIEW program with the following features.

1. Measure and store the LVDT output voltage.

2. Develop and store the standard deviation of the LVDT output voltage.

3. Apply a calibration constant to the LVDT measured voltage to arrive at a change in displacement.

4. Allow the input of the initial length of the wire.

5. Calculate, display, and log the strain the wire is undergoing.

6. Allow the input of the load (weight) being used to obtain creep deformation.

7. Calculate, display, and log the tensile stress the wire is undergoing as it is loaded.

8. Plot the strain vs. time in a plot.

9. Include the ability to modify the acquisition time and the sample rate. You will change the sample rate throughout the experiment. Do not exceed 9 s acquisition time.

10. Hints: Use a Gain subVI in a while loop. Write data out with a write to spreadsheet tool. Wire channel AI0 as the LVDT voltage. You will need a shift register for each value you wish to write out. If you wish to plot any values vs. time, you will need to create a time array. Use the plot indicator from the bottom of the Gain subVI for a quick plot, remember to bundle two single dimensional arrays to connect to the plot. Make any value that may need adjusting a control on the front panel of the LabVIEW VI. Put indicators on the raw data, channel AI0.

6.3.7   EXPERIMENTAL TASK

1. Wire the LVDT to your DAQ. Follow wiring directions supplied by the LVDT vendor.

2. Power on the LVDT. Check that you have data from the LVDT in your VI.

3. Find the calibration constant supplied with the LVDT and input it into your VI.

4. Establish the diameter of the wire.

5. Cut a piece of wire of appropriate length.

6. Attach one end of the wire to the weight hanger as pictured in Fig. 6.3. Bind one end of the wire under the thumb screw.

Figure 6.3: Illustration of wire attachment to the hanger. Attach the wire to the hanger and route the wire to outside of a post on the hanger so it aligns with the shaft of the hanger.

7. Attach the other end of the wire to the top of the tensile loading fixture, as seen in Fig. 6.4. Clamp the wire between the washers, between the thumbscrew and aluminum plate. The

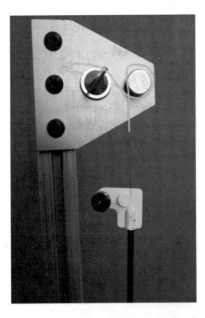

Figure 6.4: Illustration of a wire attachment to the top of a loading fixture.

wire should pass between the two binding posts. Keep the length of wire between the top and the weight carrier as long as possible. Discuss why you need long wire in your report. Measure the initial length.

8. Loosen the LVDT carrier and raise it, inserting the LVDT core that is on the bottom of the weight hanger into the LVDT body. Adjust the LVDT carrier so that the VI reads an LVDT voltage at the bottom of the device's range (Fig. 6.5), allowing for full utilization of the LVDT's travel as the wire stretches.

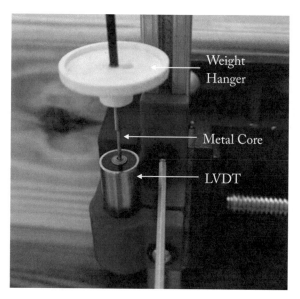

Figure 6.5: Illustration of hanger core insertion into the LVDT cylinder.

9. Steps 10–14 are examining primary creep, secondary creep, and creep recovery; two wire samples will be used. Start your VI with a sample rate of 0.1 s and a sample frequency of 100 Hz, and apply ~6 Newtons.

10. Observe the strain vs. time relationship, and when the primary creep phase is over (there will be a marked change in the rate of deflection, it will slow down), reduce the sample rate to 8 s (instructions on board). Take data for 20 m in the secondary creep phase.

11. Change the sample rate to 1 s.

12. Unload the sample and take data for 30 s.

13. Change the sample rate to 5 s and take data for 4 1/2 m.

14. Stop the program and store data.

15. Repeat steps 10–14 for the second suggested weight with creep relaxation.

16. Load another wire to examine tertiary creep. Start your VI with a sample rate of 0.1 s and a sample frequency of 100 Hz, and apply a load close to but less than the ultimate load found in Lab 3a.

17. Observe the behavior until the wire breaks; stop and store data.

18. Repeat any part of the experiment if necessary.

6.3.8   ISSUES TO BE DISCUSSED IN THE LAB REPORT

1. Develop a creep diagram for each run.

2. Show a plot of the strain vs. the time (% vs. seconds).

3. For the first two runs, indicate in your plot, the elastic strain portion, the primary creep potion, the secondary creep potion, the elastic relaxation, and the creep relaxation portion for the linear plots.

4. For the last run, note all phases of creep observed.

6.3.9   PRINCIPAL EQUIPMENT REQUIREMENTS AND SOURCING

• Multi-function DAQ with USB cable: Minimum 2 channel 14 bit ADC, USB interface, LabVIEW support. Typical vendor: Out of The Box DAQ,

  https://ootbrobotics.com/.

• LVDT: RDPE.com, DCTH400.

• 30 AWG copper wire, Vendor: McMaster Carr.

• Weight hanger: fabricated, fdm prototype and 80-20 parts.

• Fixture for holding weight hanger: fabricated.

• Calibration weights: slotted brass weights. Typical vendor: Manson Labs, others.

• Commercial scale, typical part: Symmetry EC2000 or similar.

• Measurement tools in lab (rulers, micrometers, etc.).

Charpy Impact Testing

PART A: THEORY

7.1   MOTIVATION

Resistance to impact loads is an important consideration while selecting materials for suddenly applied loads, such as those encountered during vehicle accidents, blast loads, bullet impact on glass windows or armor plate, impact by a foreign object (meteor), etc. Automobile manufacturers often test vehicles for crashworthiness and structural engineers may test structures for resistance against blast loads or explosions. Body armor should withstand the incoming kinetic energy of a bullet. The kinetic energy of impact or energy released by the explosive blast should be quickly dissipated by the material through various energy dissipation mechanisms, such as plastic deformation and fracture, so as to protect structures, precious goods, and lives. In this laboratory, we will test materials for their ability to absorb impact energy using the "Charpy" impact test method.

7.2   THEORY OF CHARPY IMPACT TESTING

The Charpy impact tester consists of an impactor mounted at one end of a metal rod whose other end is anchored to a rigid frame at a pivot point; see Fig. 7.1. The impactor and the rod serve as a simple pendulum to impose impact forces on a test sample when the pendulum is held up and left to fall freely. The specimen material (usually a beam of rectangular cross section) is placed at the lowest point of the pendulum path and is supported at two ends (similar to a simply supported beam under three-point flexural loading) such that it resists the motion of the pendulum when impacted. Upon release, the pendulum swings, breaks the specimen, and reaches a new maximum potential energy state on the other side of specimen. The difference in potential energy between the original position and the final position of the pendulum is used to quantify the energy absorbed in the specimen during the impact. Several non-conservative forces are also present, and they can be estimated and corrected for.

    The pendulum rod in the Charpy tester can be instrumented with an analog position sensor at the pivot point (bearing) to measure its angular position during the motion. The sensor generates a voltage proportional to the angular position of the impactor. For example, the output may span 5 V DC for 0°–360° rotation for a given sensor model. The 0° position (with respect to vertical) must be established by the student by freely swinging the pendulum (assuming no frictional losses at the pivot and losses due to air resistance while pendulum is in motion) and noting the extreme angular positions (voltages). The average between the initial and final position is obtained and subtracted from initial and final reading so as to establish the zero position for the pendulum. The mass of the impactor is measured ahead, and the center of mass is established for the pendulum rod with the impactor. This can be done by assuming the impactor and the rod to have regular shape and knowing their center of masses.

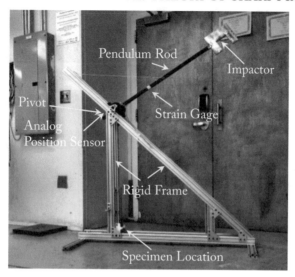

Figure 7.1: The Charpy impact tester with sensors and strain gages.

7.2.1   WIND RESISTANCE AND FRICTIONAL LOSSES

For an ideal pendulum with no losses, the initial height (H_i) should be equal to the final height (H_f) when no specimen is present, i.e., $H_i = H_f$. However, the pendulum will lose some energy to aerodynamic drag and mechanical friction in the bearings. These loses can be experimentally quantified by performing a free-swing test with no specimen present. Displace the pendulum from its minimum potential energy location to some height and allow it to freely oscillate two or more cycles while simultaneously recording data. Observe the potential energy difference between the initial position and the first zero velocity position on the other side after release. This energy loss is attributed to the non-conservative forces (friction in the bearing and wind resistance) acting on the pendulum and can be used as an approximation of losses when estimating the energy absorbed by an impacted specimen.

The energy lost during a swing (due to air resistance, friction in the pivot bearing etc.) can be found from simple geometry and kinematics of the pendulum as presented in Fig. 7.2. The initial pendulum location is noted as H_i and the location of the pendulum after it reaches the maximum height during its first swing (where it's velocity is zero) is noted as H_f. The change in potential energy is given by

$$\Delta(PE) = Mg(H_i - H_f) = \text{losses} \tag{7.1}$$

where,

- θ_i $\rightarrow$ Initial angle of impactor.

- θ_f $\rightarrow$ Angle of impactor after impact where impactor has zero velocity.

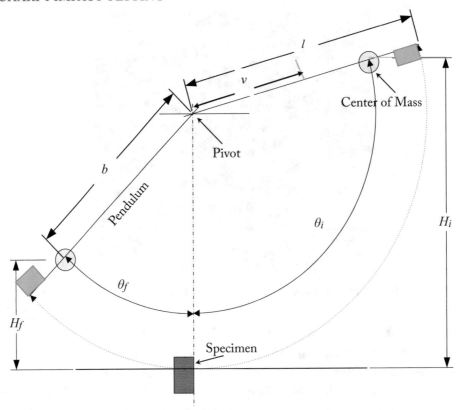

Figure 7.2: Description of variables for calculation of potential energy in the system.

- b → Effective length of pendulum (distance from pivot to center of mass).

- l → Length along impactor from pivot point to impact location.

- H_i → Initial height (function of θ_i → ground to center of mass).

- H_f → Final height (function of θ_f → ground to center of mass).

- v → Strain gage location from pivot point.

7.2.2   MONITORING OF FORCES DURING IMPACT

The pendulum is instrumented with two strain gages in 1/2-bridge configuration (on opposite sides of the beam) at some location at a distance v from the pivot. During the impact with the specimen, the pendulum bends due to the resistance offered by the specimen and the strain in the beam can be recorded and used to estimate the force acting on the specimen. This force, along with impactor position data, can be used to estimate the work done on the specimen (strain energy absorbed in breaking the specimen), allowing comparison to the change in potential energy during impact [Eq. (7.1)]. From a mechanics' perspective, the analysis for energy absorbed in the specimen involves consideration of two simply supported beams (pendulum and specimen) as shown in Fig. 7.3. Both the beams are assumed to be loaded in flexure under three-point loads. The pendulum has forces acting at the pivot point, impact point and center of mass. The specimen is resting against two end-supports and impact force (P) at the center. The forces acting on each of the two beams are shown in Fig. 7.3.

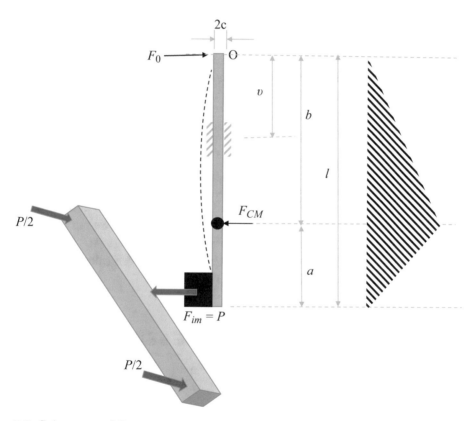

Figure 7.3: Schematic of forces acting on pendulum and specimen. The moment diagram for the pendulum is shown as a hatched triangle on the right. The deflection of the pendulum at the time of impact is shown as a dotted line.

The intent here is to estimate the impact force (F_{im}) for calculation of the energy absorbed in the specimen during the fracture. When a specimen is present, the difference in potential energy between the initial and final positions of the pendulum can be equated to sum of the energies (U) to break the specimen, air resistance, bearing resistance, and other losses.

$$mg(H_i - H_f) = U_{\text{break the specimen}} + U_{\text{air + bearing resistance + losses}} \qquad (7.2)$$

The second term on the right-hand side of Eq. (7.2) has already been calculated in Eq. (7.1). The first term on the right-hand side is the energy absorbed in breaking the specimen which involves elastic deformation of the specimen, plastic deformation of the specimen and then fracture (creation of new surfaces). From mechanics of materials, one can calculate the energies due to elastic and plastic deformation for three point bending of a beam as

$$U_{el} = \frac{P^2 l^3}{96EI} \quad \text{and} \quad U_{pl} = \frac{9P^2 L^3}{128EI} \qquad (7.3)$$

All quantities refer to the specimen being tested. Assuming negligible energy for creation of new surfaces (fracture), compare the results from Eqs. (7.2) and (7.3). However, the value of P (impact force) in Eq. (7.3) is yet unknown. It is nothing but the impact force F_{im} imparted by the pendulum on the specimen to cause fracture. To determine $P = F_{impact}$, conduct flexural analysis of the pendulum beam.

7.2.3   DETERMINATION OF F_{impact}

The pendulum (beam) has three forces acting on it: F_o at the pivot, F_{CM} at the center of mass, and F_{im} at the impact location (see Fig. 7.3). Recall that strain gages are bonded at a random location of the pendulum, at a distance v from the pivot point O. Since we can measure strain at this location during impact, we can calculate the stress at this strain gage (SG) location. Drawing a free-body diagram at the cut-section of strain gage location, we can calculate the moment at strain gage (M_{SG}) in terms of force at the pivot (F_o) as:

$$M_{SG} = F_o \times v \qquad (7.4)$$

Using Eq. (7.4), the stress at the strain gage location is calculated from flexure equation as:

$$\sigma_{SG} = \frac{M_{SG}c}{I} = \frac{F_o v c}{I} \qquad (7.5)$$

where $2c$ is the height of the cross-sectional dimension of the pendulum at the strain gage location (see Fig. 7.3) and I is the moment of inertia of its cross section.

To express F_o in terms of F_{CM}, we take moment about the impact point, i.e.,

$$F_o l = F_{CM} a \qquad (7.6)$$

where $a = (l - b)$. Inserting Eq. (7.6) into Eq. (7.5)

$$\sigma_{SG} = \frac{M_{SG}c}{I} = \frac{F_o v c}{I} = \frac{F_{CM} v c a}{Il} \qquad (7.7)$$

Since strain is measured at the strain gage location, using Hooke's law ($\sigma = E\varepsilon$), Eq. (7.7) can be written as

$$\sigma_{SG} = \frac{F_{CM} vca}{Il} = E\varepsilon \tag{7.8}$$

where E is the Young's modulus of the pendulum material. Also by taking moment about the pivot point O, we can write the force of impact F_{im} as

$$F_{im} = \frac{F_{CM} b}{l} \tag{7.9}$$

Substituting Eq. (7.8) into (7.9) gives the F_{im} as

$$F_{im} = \frac{F_{CM} b}{l} = \frac{E\varepsilon Ib}{avc} = P \tag{7.10}$$

Using this value of P in Eq. (7.3) to calculate U_{el} and U_{pl}, compare the energy for fracture in Eq. (7.2) to these values.

We can also calculate energy absorbed in another way. Note from Eq. (7.10) that F_{im} is calculated from strain measured on the pendulum as a function of time. Similarly, the angular position (θ) is also measured as a function of time during the impact process. From both these measures, one can plot F_{im} vs. θ for the interval when the impactor is in contact with the specimen. However, to calculate the energy consumed (total work done) during the impact (fracture energy), the angle must be converted to the arc length $r = l\theta$, where l is the length of the pendulum. Now plot F_{im} vs. $l\theta$ to calculate the energy absorbed during the impact. Compare this value to the values calculated in Eqs. (7.2) and (7.3). Explain the discrepancies and the accuracy of each method.

Sample Problem

1. The period of a pendulum (for small amplitudes) is given by $T = 2\pi \sqrt{(L/g)}$. A small perturbation is given to the mass and the position verses time is recorded. The pendulum swings by an angle of 5° on each side and it takes 2 s for each period. The mass of the pendulum is 20 kg and is assumed to be concentrated at the end of the pendulum rod. How many Joules are stored in the pendulum when raised to 120° from its equilibrium position?

2. Given a pendulum with a mass-less rod and an end mass of 20 kg, estimate the energy stored in the pendulum when it is raised 90° from its equilibrium position.

PART B: EXPERIMENT

7.3   CHARPY IMPACT TESTING

7.3.1   OBJECTIVE

Perform Charpy impact testing experiments on two samples: ductile and brittle material.

7.3.2   BACKGROUND

Background on Charpy Experiments: See theory notes and Relevant Standards: ANS E23–12c, ISO 14556:2000.

A schematic of the Charpy impact tester, relevant specimen, and pendulum dimensions, and a simplified VI for performing the test are shown in Figs. 7.4–7.7.

7.3.3   PRELAB QUESTION

Draw a free body diagram of the impactor pendulum and the specimen.

Explain how you would calculate the impact force and energy absorbed in the fracture of the specimen. Assume a steel rod of length 0.8 m and self-weight of 300 g supporting a hammer of 1.5 kg at the end and calculate its center of mass.

7.3.4   EQUIPMENT NEEDED

- Charpy Impact Tester (~15–25 J) with an angular position sensor and a strain gaged pendulum (see Fig. 7.4).

- Laptop computer with LabVIEW installed and functioning (provided in lab) Multifunction DAQ (provided in lab).

- Strain gage amplifier in 1/2-bridge.

- Brass notched specimen 6.5 mm × 6.5 mm × 37 mm (see Fig. 7.5).

- Marble notched specimen 10 mm × 24 mm × 48 mm (see Fig. 7.5).

- Rulers, micrometers, calipers.

7.3.5   REQUIRED LABVIEW PROGRAM (VI)

See Fig. 7.7.

7.3.6   PROBLEM STATEMENT

Examine Charpy impact results for a brass and marble specimen. Estimate the energy lost by the impact machine using a potential energy approach. Estimate the work done on the specimen

Figure 7.4: Image of a Charpy impact tester.

Figure 7.5: Image of notched Charpy impact specimens.

by evaluating the force applied over the impact of the specimen. Compare the two results for each specimen. Compensate for wind resistance and bearing friction in the potential energy approach.

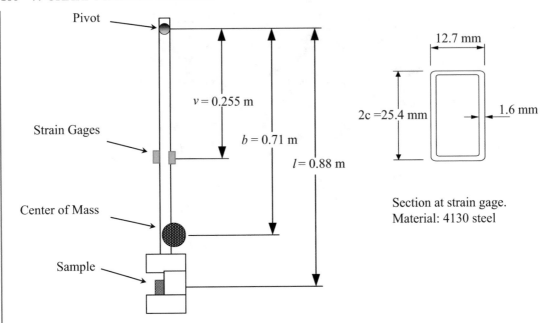

Figure 7.6: Schematic of the Charpy impact test and relevant dimensions.

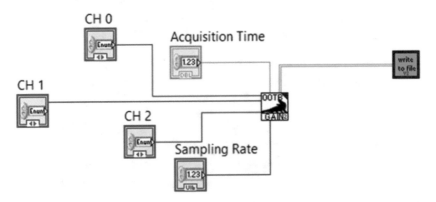

Figure 7.7: Simplified VI required to obtain data during the test.

7.3.7   EXPERIMENTAL PROCEDURE

Using the tester in the laboratory, perform the following steps.

1. Establish the 0° position: The Charpy tester has an impactor instrumented with an analog position sensor at the pivot-point that generates a voltage proportional to the angular position of the impactor relative to ground. (For exemplo, the output may span 5 V DC for 0°–360°, i.e., 72°/V). A DAQ is connected to the position sensor. Freely swing the

pendulum and note the extreme positions by recording the voltages. Average the extreme readings and subtract this value from the two extremes to establish the zero position with respect to the vertical.

2. Establish the mass of the impactor and the "center of mass" of the pendulum by knowing the weight and center of gravity of the bar and the impactor. (In the current lab the mass is 2.3 kg ± 0.05 kg and the center of mass is approximately 0.71 m from the pivot point).

3. Calculate the energy loss (due to air resistance and friction in the bearing) in the free-swing of the pendulum by noting the differences in potential energy at the two extreme positions of the swing. Displace the pendulum from its minimum potential energy location and allow it to freely oscillate two or more cycles while recording data. Observe the potential energy difference between the initial position and the first zero velocity location of the impactor after release. This energy loss is attributed to the non-conservative forces acting on the system and can be used as an approximation of losses when estimating the energy absorbed by an impact specimen. The change in potential energy is $\Delta(PE) = Mg(H_i - H_f) = $ energy loss.

4. Monitoring of forces during impact: The beam of the pendulum is instrumented with two strain gages in 1/2-bridge configuration. A strain gage amplifier is attached to the gages. The output of the amplifier is attached to a DAC. The strain in the beam can be recorded during a swing of the pendulum. Using Fig. 7.6 and the discussion provided in the Theory section (Section 7.1), as well as the strain measured, estimate the forces acting on the specimen during impact. These forces, along with impactor position data, can be used to estimate the work done on the specimen, allowing comparison to the change in potential energy for impact. The following steps should be performed during the test.

   (a) With a acquisition time set to 5 s, sampling rate 5,000 Hz, and the channels set to ± 5 V DC (subject to instrument capabilities).

   (b) Take data for position during a large (~90° displacement) perturbation swing down test.

   (c) Take 5 s of data with the impactor hanging at lowest potential energy position (0°). Use this data to establish the zero degree voltage reading for the pendulum.

   (d) Prepare a ductile sample of brass for impact, and conduct impact test. Capture pendulum position and strain gage readings with the attached DAQ.

   (e) To perform the test (requires two people, with TA/instructor):
   - Load specimen.
   - Prepare VI to run (see Fig. 7.7).
   - Lift impactor to desired height.
   - Start VI.

- Drop impactor.
- Save data.

5. Prepare a brittle sample of marble for impact and impact test. Capture pendulum position and strain gage readings with the attached DAQ.

7.3.8   ISSUES TO BE DISCUSSED IN THE LAB REPORT

1. Compute the wind resistance and friction loses from the large perturbation swing down test and develop corrections for the impact tests.

2. Calculate and report a corrected energy absorption number for the brittle and ductile samples.

3. Using the strain gage data and pendulum position data, calculate the work done during the impact of each specimen. This will require some data manipulation to extract the relevant data range from a much larger range provided.

4. Compare and discus the results from 2 and 3.

5. Present an uncertainty analysis for 2.

7.3.9   EQUIPMENT REQUIREMENTS AND SOURCING

- Multi-function DAQ with USB cable: Minimum 2 channel 14 bit ADC, USB interface, LabVIEW support. Typical vendor: Out of The Box SADI DAQ,

  https://ootbrobotics.com/

- Strain gage amplifier, typical part: Tacuna Systems EMBSGB200_2_3 configured for 1/2-bridge,

  tacunasystems.com

- Charpy impact testing apparatus: fabricated or Instron, etc.

- Charpy specimens: fabricate as needed, also Laboratory Devices Co, Inc.

- Rulers, micrometers, calipers: Various.

LABORATORY 8

Flexural Loading, Beam Deflections, and Stress Concentration

PART A: THEORY

In Lab 2, theory of cantilever beam was used to measure the stress in a beam and then weight of a beverage can. In this lab, we use a simply supported beam to study flexural stress distribution in a beam, deflections in a beam, and stress concentrations.

8.1   STRESS IN A BEAM

Consider a simply supported beam of length L, width b and height h. Stress at any point in the beam is given by

$$\sigma_x = \frac{My}{I} \tag{8.1}$$

where M is the moment at that cross section, y is the distance from neutral axis (*NA*) of the beam to the point of measurement, and I is the moment of inertia about the *NA*. For a beam with rectangular cross section, $I_{NA} = bh^3/12$. A schematic of a simply supported beam with a central load is shown in Fig. 8.1a, and a small section of the deflected beam with stress state above and below the *NA* is shown in Fig. 8.1b.

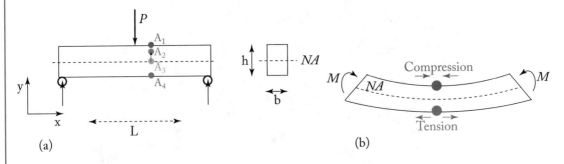

Figure 8.1: (a) Schematic of a simply supported beam and (b) a small section of the deformed beam with stress state above and below the neutral axis.

Note that even though forces are applied along y-axis in Fig. 8.1a, the stress generated is along the x-axis. For this loading, fibers of the beam below the *NA* elongate (stretch) and those above the *NA* compress along their length, as shown in Fig. 8.1b, and hence, the stress $\sigma = \sigma_x$ acts along the beam length (or x-axis). Since only one stress is developed along the axis of the beam, we can use 1D stress–strain relationship in subsequent analysis.

In Fig. 8.1a, four points are shown on the beam, A_1, A_2, A_3, and A_4. Stress at these points is

$$\sigma_{A_1} = \frac{My}{I} = \frac{M\left(\frac{h}{2}\right)}{I}$$

$$\sigma_{A_2} = \frac{My}{I}$$

$$\sigma_{A_3} = \frac{My}{I} = \frac{M\left(0\right)}{I}$$

$$\sigma_{A_4} = \frac{My}{I} = \frac{M\left(\frac{h}{2}\right)}{I}$$

However, A_1 and A_4 are on opposite sides of the beam NA. From Fig. 8.1b, it is seen that the points above the NA are in compression ($-$) and those below the NA are in tension ($+$). The y-axis is ($+$) upward (see Fig. 8.1a), which means we have positive ($+$) stress in the regions where y is negative. To account for this, we write

$$\sigma_x = -\frac{My}{I} \tag{8.2}$$

Therefore, σ_{A_1} are σ_{A_4} are opposite in sign.

From Fig. 8.1, we can see that moment M varies depending on the location of the cross section where the stress is required and hence the stress varies along the length even when y and I remain constant. To simplify calculations and get an overall picture of critical locations where stress could be maximum along the length of a beam, we draw shear-moment diagrams.

8.2 BENDING MOMENT DIAGRAM

8.2.1 SIMPLY SUPPORTED BEAM

Consider a simply supported beam with applied load at the center, as shown in Fig. 8.2a. This configuration is often referred to as three-point bending. To find stress at any location on the beam, simply make a vertical cut at that location and draw a free body diagram for that section with forces and moments. Then use equations of equilibrium ($\sum F = 0$ and $\sum M = 0$) at the cut section to find the shear force and bending moment at that location. For two cut sections, each at a distance x, the internal forces and moments are given in Fig. 8.2b and c.

In Fig. 8.2b, when the cut is made in a section $0 < x < \frac{L}{2}$, $\sum F = 0(\uparrow +)$ gives the force at the cut, i.e.,

$$\frac{P}{2} - v_x = 0 \rightarrow v_x = \frac{P}{2} \tag{8.3}$$

Similarly, $\sum M = 0(\circlearrowleft +)$ gives moment about the section as $-\frac{P}{2}(x) + M_x = 0$

$$\boxed{M_x = \frac{P}{2}(x)} \tag{8.4}$$

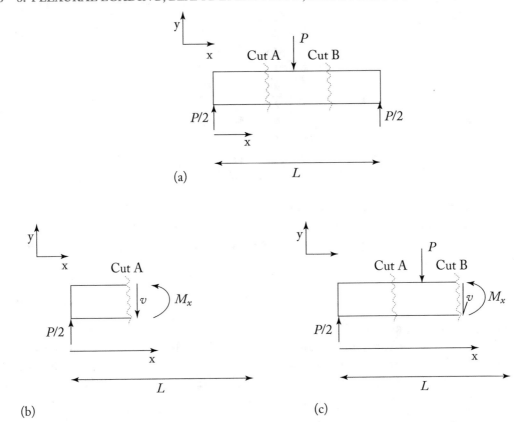

Figure 8.2: (a) A simply supported beam with two imaginary cut sections and (b) schematic indicating internal force and moment at the cut sections A and B.

Similarly, in Fig. 8.2c when a cut is made in section $\frac{L}{2} < x < L$,

$$\sum M = 0 \rightarrow -\frac{P}{2}(x) + P\left(x - \frac{L}{2}\right) + M_x = 0$$

$$\boxed{M_x = \frac{P}{2}x - P\left(x - \frac{L}{2}\right)} \qquad (8.5)$$

$$\sum F_y = 0 \rightarrow \frac{P}{2} - P - v_x = 0 \rightarrow \boxed{v_x = -\frac{P}{2}} \qquad (8.6)$$

Now, the resulting shear and moment diagrams can be drawn, as shown in Fig. 8.3.

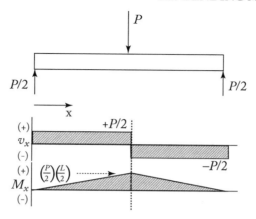

Figure 8.3: Shear and moment diagrams for a simply supported beam.

8.2.2   SIMPLY SUPPORTED BEAM WITH TWO FORCES ACTING AT EQUIDISTANT FROM END SUPPORTS

Now, let us extend the same concept for a simply supported beam with two forces acting at equidistance from the supports, as shown in Fig. 8.4. This configuration is often referred to as four-point bending. The student is encouraged to derive the shear force and bending moment equations for each segment of the beam. The shear and moment diagrams are shown in Fig. 8.4.

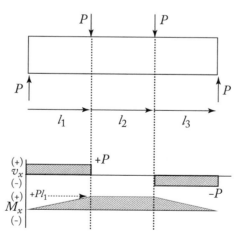

Figure 8.4: Shear and moment diagrams for a simply supported beam with two forces equidistant from end supports.

Once the moment is known at any section, the stress at any point on that cross section can be determined using Eq. (8.2).

Question:

What is the benefit of using a four-point bend beam compared to a three-point bend beam?

Recall that whenever we apply concentrated forces, stress concentration exists at these locations. In three-point bending, stress concentration exists at the center of the beam in addition to having the highest bending moment (and hence maximum stress). In four-point bending, the bending moment is maximum over the central span of the beam and there is no stress concentration in the central section (although it exists at the two load points). Therefore, while testing materials (e.g., brittle materials) whose strength is strongly dependent on internal defects (e.g., cracks, pores, impurities, second phase particles, etc.), if we use three-point bending, the failure will like occur at the center of the beam because of the combined influences of the inherent material defects and the stress concentration due to the externally applied point load. It will be unclear if the beam failed because of its inherent microstructural weaknesses or due to external stress concentration at the center of the beam. On the other hand, in four-point bending, there is no stress concentration in the central span and so the material true strength can be determined because there is higher probability that the beam will fail due to failure initiation from inherent defects. It is still possible that failure could occur at the load points. But if the failure occurs in the central span, then it must be due to the inherent defects. Thus, the true flexural strength can be determined from four-point bend better than the three-point bend test.

8.3   STRESS CONCENTRATION

Stress is always defined at a point and the formulae we use for calculating axial stress ($\sigma = F/A$), torsion ($\tau = Tc/J$), and bending ($\sigma = My/I$) provide average values because they assume a well-defined uniform geometry and uniform stress state in the calculation. However, when an abrupt change in geometry is encountered, the stress value does not scale linearly, but magnifies depending on the geometry and dimensions of the feature responsible for the abrupt change or discontinuity. Example of such features are notches, holes, cracks, etc. Stress in the vicinity of these defects is significantly higher than that in regions farther away from these features. This enhancement of stress is expressed by a stress concentration factor, k. Therefore, the max stress near a discontinuity can be expressed in flexural loading as

$$\sigma_{max} = k \frac{My}{I} \tag{8.7}$$

The k value depends on the severity of the change in geometry and the type of load applied. For design purposes, it is not important to know the exact distribution of stress around the discontinuity. Instead, the max stress should be considered.

In our Lab, we have provided a beam with two kinds of geometrical discontinuities in the beam sections where the stress varies linearly (see moment diagram of Fig. 8.4) with length:

a notch and a hole, as shown in Fig. 8.5. Strain gages are bonded on the curvature of these discontinuities to measure strain under applied loads P.

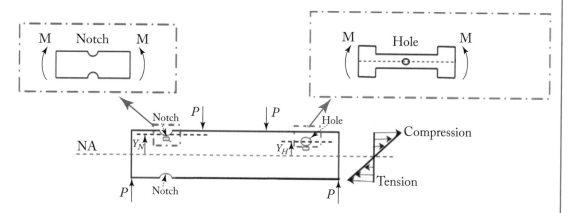

Figure 8.5: A four-point flexural beam with notches and a hole in sections where moment varies linearly.

Using stress concentration factors available from the Internet or in your Mechanics of Materials textbook, calculate the max stress at these locations and compare these values with those calculated from the strain gages measurement. Discuss the discrepancies between the two values.

For the notch or the hole, assume the beam at that location to be under pure moment on either side. The value of the moment can be calculated from considering equations of equilibrium at that section (Eqs. eqrefch8.eq4 or (8.5)) or bending moment diagram. Once again knowing the stress concentration factor (k), one can estimate the actual stress at the strain gage location which can be compared to the measured stress from the strain gage (use $\sigma = E\varepsilon$).

8.4 BEAM DEFLECTIONS

Once again, we will consider both the three-point and four-point simply supported beams discussed earlier. The intent here is determine beam deflection at the center of the span. Let us consider a simply supported beam in a centrally loaded configuration, as shown in Fig. 8.6a.

The equation for elastic curve is related to the internal moment at a cross section at distance x ($0 < x < L/2$) from the end support as per Fig. 8.6b is

$$EIy'' = M_x$$

$$EIy'' = \frac{P}{2}x$$

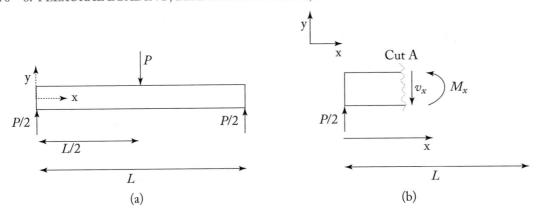

Figure 8.6: (a) A three-point simply supported beam and (b) illustration of internal forces and moments at a cross section where $0 < x < L/2$.

Integrating twice,

$$EIy' = \frac{P}{2}\left(\frac{x^2}{2}\right) + C_1 \tag{8.8}$$

$$EIy = \frac{P}{4}\left(\frac{x^3}{3}\right) + C_1 x + C_2 \tag{8.9}$$

The boundary conditions for a simply supported beam are: At the left end, $x = 0$, $y = 0$, therefore, $C_2 = 0$ from Eq. (8.9).

At $x = L/2$, the slope of the deflected (bent) beam is zero, i.e., $y' = 0$, as shown in Fig. 8.7.

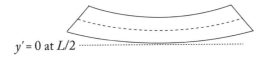

$y' = 0$ at $L/2$

Figure 8.7: Illustration of the slope of the beam at the center of a three-point simply supported.

Therefore, from Eq. (8.8), $C_1 = -\frac{1}{16}PL^2$. Hence, the deflection curve (or elastic curve) equation is given by

$$y = \frac{1}{EI}\left(\frac{P}{12}x^3 - \frac{1}{16}PL^2 x\right) = \frac{P}{48EI}\left(4x^3 - 3L^2 x\right) \tag{8.10}$$

The max deflection occurs at $x = L/2$, and thus,

$$y_{max} = \frac{P}{48EI}\left(4\left(\frac{L}{2}\right)^3 - 3L^2\left(\frac{L}{2}\right)\right) \tag{8.11}$$

$$y_{\max} = \frac{-PL^3}{48EI}.$$ (8.12)

Note: The negative sign indicates that the deflection is downward, i.e., in the negative y-direction.

Now, we extend the same approach to the four-point bend beam (Fig. 8.8a) used in the laboratory by taking cut sections at two places along the length.

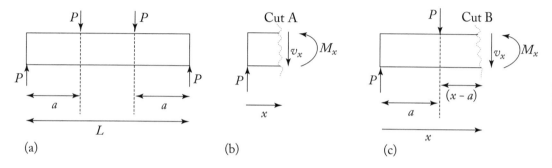

Figure 8.8: (a) A four-point simply supported beam, (b) illustration of internal forces and moments at a cross section where $0 < x < a$, and (c) at $a < x < (L - a)$.

For $0 \le x \le a$ (Fig. 8.8b)

$$EIy_1'' = Px$$

$$EIy_1' = \frac{1}{2}Px^2 + C_1$$ (8.13)

$$EIy_1 = \frac{1}{6}Px^3 + C_1x + C_2$$ (8.14)

For $a < x < (L - a)$ (Fig. 8.8c),

$$EIy_2'' = Px - P(x - a)$$
$$EIy_2'' = Pa$$

Note: The moment is constant between the inner supports.

$$EIy_2' = Pax + C_3$$ (8.15)

$$EIy_2 = \frac{P}{2}ax^2 + C_3x + C_4$$ (8.16)

Boundary conditions $x = 0$, $y = 0$, therefore, $\boxed{C_2 = 0.}$

At $x = L/2$, $y_2' = 0$; so from Eq. (8.13) and Eq. (8.15),

$$EIy_2' = Pax + C_3$$

$$0 = Pa\left(\frac{L}{2}\right) + C_3$$

$$\boxed{C_3 = -\frac{PLa}{2}}$$

Matching condition at $x = a$,

$$y_1' = y_2'$$

therefore,

$$\frac{1}{2}Px^2 + C_1 = Pax + C_3$$

$$\frac{1}{2}Pa^2 + C_1 = Pa(a) + \left(-\frac{PLa}{2}\right)$$

$$C_1 = \frac{1}{2}Pa^2 - \frac{PLa}{2}$$

$$\boxed{C_1 = \frac{Pa}{2}(a - L)}$$

At $x = a$, $y_1 = y_2$.
   To complete the solution

$$y_1 = y_2$$

$$\frac{1}{6}P(a)^3 + C_1(a) + C_2 = \frac{P}{2}a(a)^2 + C_3a + C_4$$

Substitute C_1, C_2, and C_3

$$\frac{1}{6}Pa^3 + \left(\frac{Pa}{2}(a - L)\right)a(0) = \frac{P}{2}a^3 + \left(-\frac{PLa}{2}\right)a + C_4$$

$$\frac{1}{6}Pa^3 + \frac{Pa^3}{2} - \frac{PLa^2}{2} = \frac{Pa^3}{2} - \frac{PLa^2}{2} + C_4$$

$$\frac{1}{6}Pa^3 = C_4$$

$$\boxed{C_4 = \frac{Pa^3}{6}}$$

Thus, the max deflection, which occurs at $y_2 \left(x = \frac{L}{2} \right)$, is

$$y_2 = \frac{1}{EI} \left(\frac{P}{2} a x^2 + C_3 x + C_4 \right)$$

$$y_2 = \frac{1}{EI} \left(\frac{P}{2} a \left(\frac{L}{2} \right)^2 + \left(-\frac{PLa}{2} \right) \left(\frac{L}{2} \right) + \left(\frac{Pa^3}{6} \right) \right)$$

$$\delta_{max} = y_2 = \frac{1}{EI} \left(\frac{PL^2 a}{8} - \frac{PL^2 a}{4} + \frac{Pa^3}{6} \right)$$

$$= \frac{1}{EI} \left(-\frac{PL^2 a}{8} + \frac{Pa^3}{6} \right)$$

$$\boxed{\delta_{max} = \frac{Pa}{2EI} \left(\frac{a^2}{3} - \frac{L^2}{4} \right)}$$

If $a = \frac{L}{3}$, then

$$\delta_{max} = \frac{P \left(\frac{L}{3} \right)}{2EI} \left(\frac{1}{3} \left(\frac{L}{3} \right)^2 - \frac{L^2}{4} \right)$$

$$= \frac{PL}{6EI} \left(\frac{L^2}{27} - \frac{L^2}{4} \right)$$

$$= \frac{PL^3}{6EI} \left(\frac{1}{27} - \frac{1}{4} \right)$$

$$= \frac{PL^3}{6EI} \left(-\frac{23}{108} \right)$$

$$\delta_{max} \approx -0.0355 * \frac{PL^2}{EI}$$

Recall that for a simply supported three-point bend beam of the same length

$$\delta_{max} = \frac{-PL^3}{48EI}$$

So the ratio of max displacement between the two is

$$\frac{\delta_{max}^{3pt}}{\delta_{max}^{4pt}} = \frac{\left(-\frac{1}{48} \right) \frac{PL^3}{EI}}{\left(-\frac{23}{108} \right) \frac{PL^3}{6EI}} = \frac{27}{46} \approx 0.587$$

$$\delta_{max}^{3pt} = 0.587 \delta_{max}^{4pt}$$

Thus, the three-point bend beam has lower maximum deflection at the center of the span than a four-point bend beam for the same load P and for the support span lengths shown above. Students should rationalize if this result is reasonable.

Sample Problem

A simply-supported long rectangular steel ($E = 200$ GPa) beam, similar to the one shown in Figs. 8.4 or 8.9, is bonded with stain gages at four location marked A, B, C and D. The beam is of length 1.5 m, height 120 mm and thickness 5 mm. The outer support span is 1.5 m and the inner loading span is 0.5 m. Strain gages A and B are bonded on the top surface at distances 0.3 m and 0.6 m from the left end-support. Strain gage C is bonded on the front surface at 0.8 m from the left end-support and at 25 mm from the top surface. Strain gage D is bonded at 1.1 m from the left end-support and at 60 mm from the top surface.

(a) Write equations for moment and shear force on a cross section at the locations of the strain gage.

(b) For $P = 1,000\,N$, calculate the stress generated at all the locations of the strain gages.

(c) What is the strain measured by the gage at B and D?

(d) What is the deflection of the midpoint of the beam?

PART B: EXPERIMENT

8.5 MEASUREMENT OF STRESS, DEFLECTION, AND STRESS CONCENTRATION

This is an open-ended laboratory.

8.5.1 OBJECTIVE

Explore four-point bending of regular rectangular beams with introduced defects (stress concentrators). Because this lab is assigned open ended, no procedure or discussion items are provided. The student is to examine the provided apparatus and develop a thesis around four-point bending of a beam. The student is to then develop a procedure to test his thesis, and report the results of the work in a report. The student must have the thesis approved by the lab instructor before conducting the experiment.

8.5.2 BACKGROUND REQUIRED FOR CONDUCTING THE LAB

Familiarity with LabVIEW programming is essential for all the labs. Useful LabVIEW instructional videos can be found on the National Instruments website. Basic understanding of four-point bending, along with understanding of strain gage use and stress concentrations is required. Familiarity with finite element analysis (FEA) modeling of four-point bending is desired. Student is encouraged to use any commercial finite element package for analysis and comparison of the results obtained experimentally.

8.5.3 EQUIPMENT AND RESOURCES NEEDED

- Laptop computer with LabVIEW installed.

- Multi-function DAQ with USB cable.

- Four-point bending apparatus including load cell and load jack (Fig. 8.9).

- Regular rectangular beam with introduced defects and strain gage instrumentation

- 12″ adjustable wrench for applying load.

- Strain gage amps and strain gages as needed.

- Measurement tools in lab (rulers, micrometers, etc.).

8.5.4 FOUR-POINT BENDING APPARATUS WITH INSTRUMENTED BEAM

The apparatus, shown in Fig. 8.9, is designed to provide equal load at four points along a long rectangular beam that is held in a rigid frame. The outer two load points are on a rigidly anchored

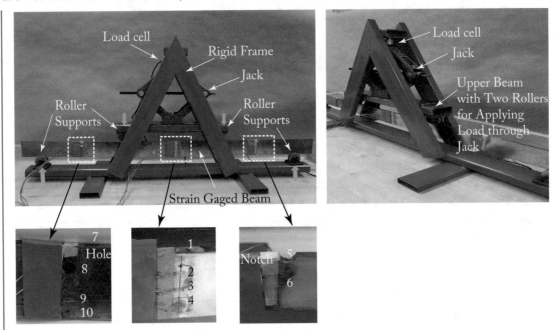

Figure 8.9: Four-point bending apparatus with an instrumented rectangular beam. The images in the magnified view show strain gage locations and discontinuities.

bottom beam and the two inner load points are anchored on another rigid upper beam. A jack applies load at the center of the upper beam. A wrench can be used to apply the desired load. At the top of the jack, a load cell is rigidly positioned to measure the applied load.

A rectangular beam of suitable dimensions is bonded with several strain gages to obtain a stress distribution along its width. A notch on the top surface and a hole mid-way between the neutral axis and top surface are drilled as shown in the figure. Strain gages are bonded close to these discontinuities to determine stress concentration factors. Depending on the breadth of analysis the student wishes to undertake, a beam without these stress concentrators may also be used. However, the student may choose to use finite element analysis to support the experimentally determined stress distribution. The locations of the strain gages are also shown in magnified views in Fig. 8.9 and tabulated below. Note that strain gages 1 and 7 are bonded on top surface of the beam (along the thickness, i.e., farthest from the neutral axis). Strain gage 5 is bonded inside the notch to get better strain reading closest to the notch, as shown in the magnified view of insets in Fig. 8.9. Similarly, a smaller strain gage is bonded inside the hole. Two additional strain gages (9 and 10) are bonded below the holes. The relevant details of the components are given in Table 8.1. Intentional redundancy in strain gage locations is built into this arrangement.

Beam: Approximate dimensions are 76 mm (3.0 in) tall, 6.4 mm (0.25 in) thick, and distance between supports ~1220 mm (48 in). These values may differ depending on the equipment and the student is urged to make these measurements. Some specimens have stress concentrations (symmetric notches and holes, all radii 6.4 mm or 0.25″). 1/4-bridge strain gages are located at points of interest. Central span of the beam (where the two load points are situated) is ~445 mm (17.5 in). The beams with stress concentrations are fabricated in O-1 tool steel, annealed.

Jack: 1,363.6 kgf (3,000 lbf) limit.

Load Cell: 1,000 kgf limit, an information sheet for appropriate load cell is provided in the lab. A "rated Output" is given (mV/V), assume this is for the rated load. The load cell used is based on a full W-bridge.

Table 8.1: Gages, locations, and distances

Gage #	Location	Distance from Top of Beam, mm (inches)	Distance from Center of Beam, mm (inches)
1	Center of the beam, on top surface	0.0	0.0
2	Center of the beam	19.5 (0.75)	0.0
3	Center of the beam	38.1 (1.50)	0.0
4	Center of the beam	57.15 (2.25)	0.0
5	Inside the notch	6.35 (0.25)	387.4 (15.25)
6	Below the notch	19.5 (0.75)	387.4 (15.25)
7	Near the hole, on top surface	0.0	387.4 (15.25)
8	In side radius of the hole	22.23 (1.125)	387.4 (15.25)
9	Below the hole	50.80 (2.000)	387.4 (15.25)
10	Below the hole	68.58 (2.700)	387.4 (15.25)

All gages have a gage factor of 2.02. Strain gage locations are accurate within $\pm$ 0.762 mm (0.03 in).

8.5.5   TYPICAL WIRING FOR STRAIN GAGES AND LOAD CELL

Each strain gage is connected to a single amplifier. The load cell is also wired through the amplifier. All the amplifiers can be run by a single power supply, as shown in Fig. 8.10. The signals from the amplifiers and the load cell are run through a DAQ to the computer.

Figure 8.10: Wire connections between strain gages, power supply, amplifiers and DAQ.

8.6 DEVELOPMENT OF LAB GOALS AND PROCEDURE

8.6.1 OBJECTIVE

To examine the four-point bending apparatus and beams, develop a thesis concerning four-point bending of the beam, and express a procedure to support the developed thesis.

8.6.2 WHY ARE WE DOING THIS?

This lab work is designed to allow the student to explore four-point bending of a beam with and/or without stress concentrations. It tests the student's ability to generate a thesis and develop an experimental process to prove or disprove the thesis.

8.6.3 CONNECTIONS REQUIRED

- Each strain gage will require a strain gage amp. The amplifier must be powered, and an excitation voltage and an amplified gage voltage must be captured by a DAQ device.

8.6.4 REQUIRED LABVIEW PROGRAM (VI)

- Write a LabVIEW program that uses the "Gain.vi" as a SubVI that captures the excitation voltage and the amplified bridge voltage associated with each strain gage considered and the load cell. Recall that the DAQ device used has limited number of analog inputs and a procedure should be developed to account for this. The VI must save the raw data gathered. Further processing of the data into strains and stresses is at the discretion of the student. The load cell data must be processed to the point that overloading of the load cell or jack can be prevented.

8.6.5   INSTRUCTIONS

- Once the DAQ connections and LabVIEW VI are functional, follow the steps developed to generate the data necessary to support the thesis.

- Ensure that the beam loading is low enough to prohibit plastic deformation of the beam.

- Ensure that the beam loading is low enough to prohibit damage to the load cell or jack.

8.6.6   ISSUES TO BE DISCUSSED IN THE LAB REPORT

During the initial parts of this lab work, develop items to discuss. Use these items to drive the development of the procedure used to verify the thesis.

8.6.7   EQUIPMENT REQUIREMENTS AND SOURCING

- Multi-function DAQ with USB cable: Minimum 2 channel 14 bit ADC, USB interface, LabVIEW support. Typical vendor: Out of The Box SADI DAQ,

  `https://ootbrobotics.com/`

- Four-point loading frame, fabricated structural steel.

- 1,000 kgf load cell, Phidgits Button Load Cell (0–1,000 kg) - CZL204 or similar.

- 3,000 lb scissor jack, amazon, or similar.

- 76.2 mm × 635 mm × 1,321 mm (3″ × .25″ × 52″) beam, material: various (O1 tool steel, 6061 T6 Al), Mcmaster Carr.

- Uniaxial strain gages, 120 Ω, Vishay or Omega

- 305 mm (12″) Adjustable wrench: Various.

- Strain gage amplifier, typical part: Tacuna Systems EMBSGB200_2_3 configured for 1/4-bridge.

- Measurement tools in lab (rulers, micrometers, etc.).

LABORATORY 9

Wave Propagation in Elastic Solids and Dynamic Testing of Materials

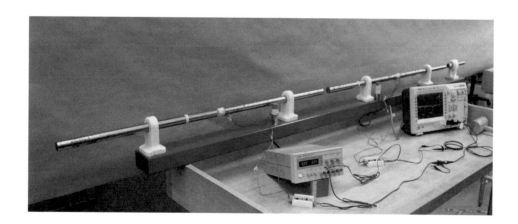

PART A: THEORY

9.1   MOTIVATION

Do you know how a scientist determines the epicenter of an earthquake? How do geologists determine the location of gas or oil deposits in the earth? How do we know that the earth crust contains molten iron? How do scientists estimate the arrival of a tsunami on to the shore? In all these cases waves play a critical role. In this lab you will learn types of waves, how fast they propagate, their interaction with a boundary, and how to use them to determine the properties of a material? For example, why does silly putty (a polymer) elongates by several folds its original length when pulled slowly but snaps when pulled fast?

   This lab consists of two parts. First, the theory of wave propagation in a slender rod is presented. Second, using this theory, the dynamic stress–strain response of a ductile metal and a brittle ceramic are determined.

9.2   BASIC CONCEPTS OF WAVE PROPAGATION

Waves are omnipresent: sound waves, light waves, electro-magnetic waves, gravitational waves, waves in ocean, etc. Verbal communication and noises around us are examples of sound waves traveling through air. Tsunami is an example of disturbances initiated somewhere in the ocean (due to an earth quake) and travels long distances before it reaches a shore. Earthquake motion is felt at distances far away from the epicenter because the disturbance travels through earth in the form of vibrations or waves. A train moving on tracks can be heard from a distance by listening to the sound waves traveling through the steal tracks. The recently discovered gravitational waves are the disturbances generated when two black holes collided thousands of light years away. Thus, we are constantly exposed to waves. In this laboratory we will learn 1D wave motion as it applies to solids. Basic concepts about normal and shear waves will be presented and experiments will be conducted to learn their propagation characteristics.

   To illustrate this basic concept, let us consider waves in a string. When a wave is generated at one end of a string by a rapid motion of hand, it just does not remain at one location, but travels along the string. While the motion of the wave is along the length of the string, the motion of individual particles in the string is perpendicular to the length of the string. Note that it is only the wave that travels from one end of the string to the other end, but the particles of the string traverse only a small distance up and down at the same location. Here the particle motion is perpendicular to the wave motion. The wave can be thought of as a "pocket of energy" which travels along the length of the string but the particles go back to their initial position immediately after the wave passes by. Also, only the particles within the wave are in motion, and the particles outside of this wave in the string are stationary until the wave arrives and go back to that state after the wave passes by. Figure 9.1 illustrates these concepts.

   When an object strikes a surface (e.g., a hammer blow on a flat surface) a number of different types of waves are generated carrying the energy of the blow into the bulk material at

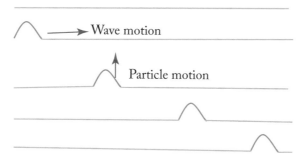

Figure 9.1: Illustration of wave propagation in a string. Note that only the wave propagates from one location to another but the particles of the string move up and down when the wave arrives. Also, only particles within the wave are in motion and the rest of the string is stationary.

different velocities and intensities. For simplicity, we will only deal with two types of waves in this class: longitudinal waves and shear waves.

1. Longitudinal waves (also called Primary waves or P-waves): In these waves, the wave motion and the particle motion are parallel to each other. They can be generated when a long rod is impacted axially by another short rod. The wave travels along the length of the rod and the particles encompassed by the wave within the rod also move in that direction. These waves travel at a velocity equal to

$$C_L = \sqrt{\frac{E}{\rho}} \qquad (9.1)$$

where E is Young's Modulus and ρ is density of the material (medium) in which the wave travels. C_L stands for longitudinal wave. These waves travel fastest of all the waves generated in a body. There are two kinds of longitudinal waves.

(a) *Compression wave:* Here the wave motion and particle motion in the material are in the same direction causes pushing (compression) of the adjacent particles as the wave moves forward.

The compressive wave action can be visualized by assuming a set of people (particles) standing next to each other (side by side) and the first person being pushed gently (particle motion). This person (particle) transfers that energy to the next person (particle) and so on. Note that only the motion or energy transfers from one person to the next but the person does not move to a different location from the original location. Also, because the person is being pushed, he/she experiences compression due to the presence of the next person who offers resistance, and then the second person receives the motion from the previous person. The first person now goes back to his or her original rest position after the transfer process is complete. This process continues from one person to the next but the position of the persons does not change although the motion has transferred from the first person to the last. Thus, the wave has moved over to a different location, but the particles (persons) remain at the same position except that each person has moved a bit in the direction of the wave when the motion arrived.

(b) *Tension wave:* Here the wave motion and particle motion are in opposite directions causing pulling (tension) of the adjacent particles as the wave moves forward. Consider once again a set of people standing next to each other. But this time they are holding each other hands at their elbows. When the first person (particle) is pulled, the person moves in that direction, but because he/she is holding the next person, the 2nd person also now moves in the same direction as the 1st person. But the wave has moved in the opposite direction, i.e., from 1st person to the 2nd. Thus, when a tensile wave arrives, the particle moves toward the wave but the wave travels in opposite direction. Thus, the particle motion and wave motion are in opposite directions.

In both compression and tension waves, the wave velocity is given by $C_L = \sqrt{\frac{E}{\rho}}$.

2. Shear waves (also called secondary waves or S-waves, rotational wave, etc.): In these waves, the wave motion and the particle motion are perpendicular to each other, i.e., assume that the two adjacent particles rub against each other (vertical motion) as the wave moves from one particle to the next. These waves can be generated when a long rod is twisted at one-end over a short length and the twist is suddenly released. Then a rotational wave travels along the length of the rod causing local twist of the rod on a plane perpendicular to the axis of the rod. The wave travels along the length of the rod but the particles of the rod move in a cross-sectional on a plane perpendicular to the axis of the rod. The shear waves travel at a velocity equal to $C_S = \sqrt{\frac{\mu}{\rho}}$, where μ is shear modulus of the material.

Let us calculate wave velocities in some solids and fluids. For steel, $E = 200$ GPa and $\rho = 7,800$ kg/m^3, and therefore, the longitudinal wave velocity (C_L) is 5,064 m/s. Similarly, knowing that $\mu = E/(2(1 + v))$, the shear wave velocity (C_s) can be calculated as 3,222 m/s. Note that many Newtonian fluids (e.g., water, air) do not resist shear and hence have no shear wave velocity. However, fluids resist hydrostatic pressure which can be measured in terms of their bulk modulus (K). For water, bulk modulus is 2.2 GPa, and hence wave velocity in water is $\sqrt{\frac{K}{\rho}}$ which is 1480 m/s. Similarly, sound wave velocity in air is 330 m/s. For comparison purposes, the velocity of light is 3×10^8 m/s.

When a stress wave travels from one material to another across a boundary, its magnitude may change. In 1D stress wave propagation, an important characteristic of a material that dictates wave interaction with a boundary or an interface is called "impedance" (I) which is defined as the product of density (ρ), cross sectional area (A), and wave velocity (C), i.e.,

$$I = \rho A C \tag{9.2}$$

The importance of impedance in wave propagation will be illustrated throughout this section.

9.3   1D STRESS WAVE PROPAGATION IN A SLENDER ROD

Propagation of waves in a 3D solid is a complex phenomenon. Therefore, in this lab, a simplified illustration using 1D stress wave propagation characteristics in a slender rod will be studied. Consider a long cylindrical rod (length to diameter ratio > 10) of length L_{in}, called the incident bar, being impacted co-axially by a "striker rod" of equal diameter but of significantly shorter length (L_s). The impact velocity of the striker rod is V_s. The subsequent wave generation process in both rods is described through a series of time-lapse diagrams in Fig. 9.2. First and foremost, student should realize that we are NOT dealing with rigid body dynamics which you have learned in your Dynamics course. In rigid body dynamics, the material is assumed to have infinite stiffness ($E \to \infty$), and therefore the wave velocity is infinity. Hence, upon impact, the entire body (every particles) is assumed to achieve the same infinite velocity instantaneously and therefore, all particles in the body move as one unit. In reality, every material has a finite stiffness and hence when it is impacted, a stress wave of finite amplitude is generated and propagates in the body, just like the waves illustrated in a rope (Fig. 9.1), except that in a metal rod the waves travel at significantly higher velocity (~5,000 m/s). For simplicity, let us assume that both rods are made of same elastic material and have same cross sectional area (i.e., equal impedance).

Upon impact by the striker rod on the incident rod, a stress wave of equal amplitude is generated in both the rods (because they have equal impedances). Initially, only the material points along the impact surfaces move in opposite directions in each body, initiating the stress wave. Rest of the particles (away from the impact surfaces) in both bodies are at rest. As long as the bodies are in contact, the stress wave continues to generate and spread into each body, going in opposite directions. Recall that the impact generates a compression wave in both rods (similar to deformation when two people collide) and hence the waves generated are compressive longitudinal waves. In these waves the wave motion and particle motion are in the same direction, right-ward in the incident bar and left-ward in the striker bar, shown in Fig. 9.2. Thus, the generated waves travel at a velocity $C_L = \sqrt{\frac{E}{\rho}}$ in both rods. The wave generation process continues until the wave in the striker bar reaches the other side of the bar which is a free-end (no stress on this surface). To keep this end-surface stress-free, the compressive wave has to convert into tension wave of equal magnitude so that the signs cancel off resulting in zero stress on the surface. This reflected tensile wave now travels back in opposite direction (toward the impact-end), thus nullifying the existing compressive stress as it moves toward the impact-end of the striker rod. In a tension wave the particles move in opposite direction to the wave motion. So, when the

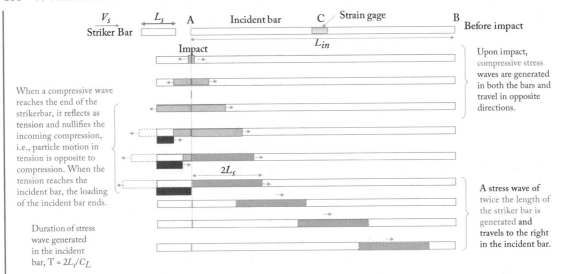

Figure 9.2: Illustration of compressive wave generation in the incident bar upon impact by a striker bar. Grey is compression wave and red is tension wave. The length of the stress wave is twice that of the striker bar.

tension wave reaches the impact end, the particles on this surface move away from the interface thus nullify (stop) the stress being applied on the incident bar. Thus, the entire kinetic energy of the striker rod has been transferred into the long (incident) rod over this contact period. During this time, the wave in the long incident rod continues to grow at the same velocity as in the striker rod. Because the wave in the striker rod has travelled twice the length of the striker rod before it stopped exerting pressure, the wave generated in the incident rod is now twice the striker rod length in physical dimension. The arrival of the reflected tensile wave in the striker bar to the impact-end signals the end of the compression stress wave generated in the incident rod. Thus, the duration (T) of the compressive stress wave generated in the long rod is equal to twice the wave travel time in the striker rod, i.e.,

$$\text{Duration of stress wave,} \quad T = \frac{2L_S}{C_L} \tag{9.3}$$

The physical length of the compression stress wave in the long rod is twice the length of the striker rod. This wave now travels in the incident rod and reaches the other end B. The above sequence of events are illustrated in Fig. 9.2 where the dotted lines at the end of the striker rod indicate the portion of the compression wave that is reflected back into the rod as a tension wave. Note that during this entire process, the rest of the particles in the long rod outside of the compression wave are at rest and do not move at all (unlike in rigid body dynamics). Only the particles in the compression wave are in motion. Once the stress wave reaches the end B, which is stress free, the compression wave reflects back as a tension wave to maintain this stress-free condition. But, before we describe this reflection process, one must also realize that the nature of reflected wave now depends on the boundary conditions of the bar, which is described in Fig. 9.2.

9.4   WAVE REFLECTION AT A FREE-END

Consider the scenario where the incident bar-end is stress free (Fig. 9.2), and hence the particles on the end surface are allowed to move freely. When the compression wave arrives at the free-end of the bar, the incoming compression wave reflects back as a tensile wave with same magnitude to maintain the zero stress condition. This process is illustrated in Fig. 9.3 with a sequence of illustrations. Recall that the direction of particle motion in a compression wave, (i.e., stress free) is in the same direction as the wave motion, and it is in opposite direction to the direction of tensile wave propagation. So, when a compressive wave reflects back as a tension wave at the free-end, going in opposite direction, the particle velocity doubles on the free surface but the stress is zero. Thus, the particles on the free surface at the end of the bar, have double the displacement compared to the particles in the interior of the bar, when the stress wave arrives. Thus, at a free-end, the particle velocity doubles, but the stress is zero because the incoming wave changes sign (compression reflects as tension and tension reflects as compression). This process is illustrated in Fig. 9.3 where the dotted lines at the end of the rod indicate the portion of the compression wave that is reflected back into the bar as a tension wave.

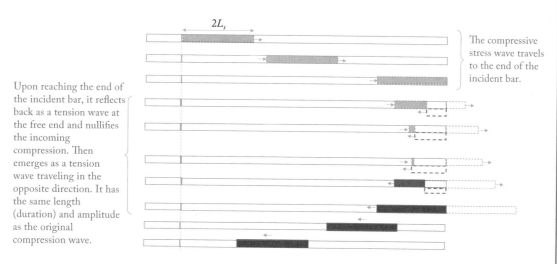

Figure 9.3: Illustration of compressive wave reflection at the free-end of the incident bar as a tension wave. The end surface is stress free, but the particles on the end surface have twice the displacement (velocity). Grey is compression wave and red is tension wave.

9.5   WAVE REFLECTION AT A FIXED-END (RIGID)

Consider a case where the bar is held against a rigid boundary and is not allowed to move. When the compression wave arrives at this boundary, because of the end constraint, the *compression wave reflects back as a compression wave* only, thus ensuring that the particles at the end surface remain at the same location because they are against a rigid (fixed) boundary. This can also be interpreted as the particles movement in the incoming compression wave are nullified by the particle movement in opposite direction in the reflected compression wave going back (away from the fixed boundary). However, the amplitudes of the incoming and outgoing compression waves remains the same and hence the total amplitude of stress at the fixed-end is twice that of the incoming stress amplitude. Thus, in summary, when the longitudinal wave encounters a fixed boundary, it reflects back without changing its amplitude or sign, i.e., compression wave reflects as compression wave only, and the tension wave reflects back as tensile wave only. Because incoming compression wave reflects back as compression wave only at the fixed-end, the stress is doubled at the fixed-end, but the particle velocity (or displacement) is zero. Figure 9.4 illustrates the sequence of events during the wave interaction at a fixed boundary.

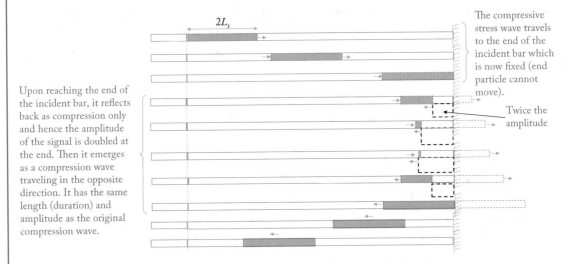

Figure 9.4: Illustration of compressive wave reflection as a compressive wave at a fixed boundary of the incident bar. Note that the stress doubles at the end surface but the displacement is zero at the fixed-end boundary.

9.6   MEASUREMENT OF STRESS WAVE DURATION AND AMPLITUDE

In the previous section we emphasized theory of wave propagation. Now, we try to experimentally study the wave propagation in the laboratory. Let us revisit Fig. 9.2 and notice that a strain gage is located at the center of the long rod. The strain gage measures the stress wave as it passes by. Note that the gage is far away from the impact end. Therefore, the gage does not detect any strain immediately upon impact by the striker but senses strain in the incident bar only when the stress wave arrives to its location. The strain gage is deformed (senses strain) for the duration $T = \frac{2L_S}{C_L}$ of the stress wave, and upon complete passage of the wave, the strain goes back to zero. This phenomenon is similar to a person sitting in a sports stadium and watching people do the "wave." Only when the wave arrives, the person stands and swings his/her hands up in the air indicating the wave arrival (measure strain). Upon passage of the wave the person goes back to the sitting position (no strain). Thus, the strain gage measures strain induced by the incoming stress wave only for the duration of the wave it is present at that location as indicated in amplitude vs. time diagram in Fig. 9.5. Similarly, the stress wave will be detected once again when it returns back from the other end. The time interval between these two measured signals is equal to the travel time for the wave to traverse to the free-end and back. Thus, by measuring the distance and time between the wave arrivals, we can calculate the wave velocity in the bar. This velocity can be compared to the theoretical estimate based on Young's modulus and density (Eq. (9.1)). Also, note that the sign of the reflected wave is reversed, consistent with the nature of reflected tensile wave from a free-end. Similarly, if amplitude-time plot is drawn for Fig. 9.4, the second wave will have the same sign as the first wave (both in compression) due to the fixed-end of the long rod. A typical experimental signal measured on a long rod for three passages of the wave is shown in Fig. 9.5a and experimentally obtained signal for three reflections is shown in Fig. 9.5b.

9.7   WAVE TRANSFER THROUGH A BOUNDARY BETWEEN TWO SIMILAR RODS

Let us now consider wave propagation across a joint (interface) between two long rods of similar material and geometry (cross-section), as shown in Fig. 9.6. We call the first bar an "incident bar" and the second bar "transmission bar." Assume that the contact surfaces of the two bars are perfectly aligned, i.e., the surfaces are smooth and perpendicular to the axis of each rod and the axes of two rods are well aligned. Also assume that a compressive wave has been generated at one end of the first rod propagating toward the other end (joint). Recall that in a compressive wave, the particles move in the same direction as the stress wave. So, when the compressive wave reaches the joint, it pushes the particles on the end-surface in the first rod toward the second rod. This action causes the wave (energy) to transfer to the second rod without any loss in magnitude or duration. Upon complete transfer, this stress wave in the second rod travels along its length

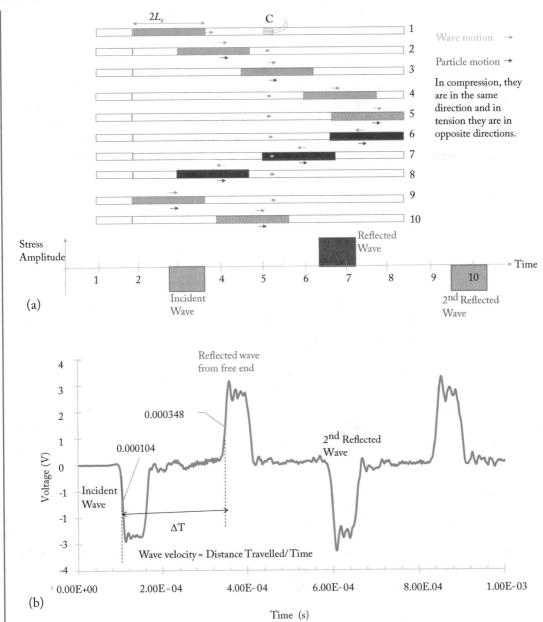

Figure 9.5: (a) Illustration of wave propagation in the incident bar, reflection at the free-end and schematic wave amplitude vs. time plot (bottom). The direction of wave motion is indicated by blue arrows and particle motion by red arrows. (b) Experimental wave profiles in an incident bar due to impact by a striker bar.

to its other end (free-end) and reflects back as a tensile wave and comes back to the joint. Will this wave now transfer fully back into the first bar? The answer is "no." Recall that the particle motion in a tensile wave is in the opposite direction to the wave motion. So, when this tensile wave arrives at the joint, the particles of the second rod at the interface move away from the boundary and cause a small gap. Thus, the entire wave now reflects back into the second rod as a compressive wave and no part of the wave transfers into the first rod. This wave, upon reaching the free-end, reflects back again as a tensile wave. This process repeats until the wave finally dies off. Thus, the energy of the compressive wave, initially generated in the first rod, is now trapped in the second rod and the wave reflects back and forth only in the second rod as the compression wave while traveling toward the free-end and a tension wave while traveling toward the joint. Eventually, the second bar will move away from the first rod due to these particle motions as in a caterpillar. The student should realize that particles within the stress wave in the second rod will always be moving to the right (towards the free-end) regardless of the wave direction whether the wave is compressive or tensile. The signals detected at the strain gages bonded at the center of each bar are shown in the Amplitude vs. Time graph in Fig. 9.6. Experimentally generated strain gage signals in two long rods are shown in Fig. 9.7 for the two scenarios discussed above. Note that the stress wave duration is less than hundred microseconds and the time required for wave to travel from the strain gage to end of the bar and back is only few hundred microseconds. From both these measurements one can calculate the stress wave velocity in the bar (exercise to be performed in the laboratory).

Now, let us summarize the concepts we have learned so far, on the wave propagation in a slender rod.

- Longitudinal wave velocity in a 1D rod is given by $\sqrt{\dfrac{E}{\rho}}$.

- In a compression wave both wave motion and particle motion are in the same direction and in tension wave they are in opposite direction.

- When a short striker bar impacts a long slender rod (incident bar), the impact generates in the incident bar a compression stress wave of length twice the length of the striker bar.

- When the incident and transmission bars are in contact, the incoming compressive stress wave remains unaltered as it crosses the interface.

- Upon reaching the end of the transmission bar it reflects back as a tension wave so as to satisfy stress-free boundary condition of the end surface.

- When the tension wave reaches the interface of the incident-transmission bars, it reflects back as a compression wave because tension cannot cross the interface that is not bonded.

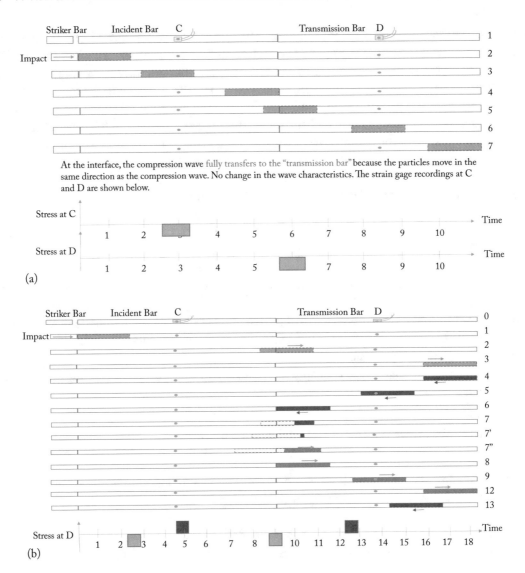

Figure 9.6: (a) Illustration of compression wave transfer across a boundary when the two rods are in contact. The compression wave transfers into the transmission bar without change in the wave characteristics. The strain gage recordings at C and D are shown below each illustration. (b) Reflection of compression wave at the free-end of the transmission bar as a tension wave and then it reflects back as compression wave at the boundary (interface) between the two rods. A compression wave can transfer across a boundary, but a tension wave cannot transfer across a boundary.

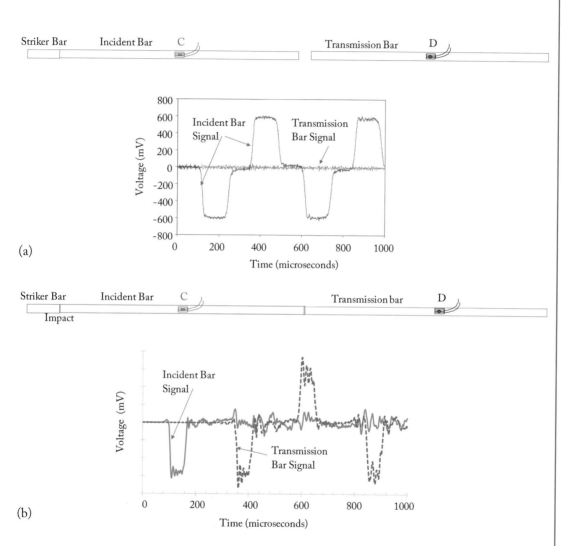

Figure 9.7: (a) Incident and transmission bar strain gage signals generated when the bars are not in contact with each other and (b) the bars are in contact with each other.

- At a free surface the incoming stress wave changes sign as it reflects back (i.e., compression becomes tension and tension becomes compression) and at fixed surface the incoming stress wave sign remains the same as it reflects back.

9.8   DYNAMIC STRESS–STRAIN RESPONSE OF MATERIALS

In Lab 3, we determined the stress–strain response of several materials when load was applied slowly over several seconds and the deformation (strain) was measured. This slow loading rate is referred to as "quasistatic loading," which means the specimen is under static equilibrium at any instant but the load is slowly increasing such that static equilibrium is still maintained. Under dynamic loads, when stress waves are used to apply loads, the entire body is not under the same level of stress but stress may exist at only one location and no stress at all other locations (unlike the tension test you have performed earlier in Lab 3). But the material within the stress wave does maintain same level of stress at all the points. The question is, do materials respond differently if the load was applied fast? The answer: Many materials do! For example, biological materials and polymers become stiffer and behave differently when load is applied fast compared to their deformation behavior when load is applied slowly. Another example is silly putty which stretches several times its original length (several hundred % strain or highly ductile) when pulled slowly but snaps (brittle behavior) when pulled fast. It exhibits a transition from ductile-to-brittle deformation behavior when rate of loading is increased. Another good example is your skin. When pressed slowly, it deforms but when slapped, you feel burning sensation. Thus, the sensors in your skin are sensitive to rate of loading. Many engineering materials behave in a similar manner, i.e., they exhibit different stress–strain behavior under dynamic loads which is the focus of this lab.

When load is applied fast (in a short duration), the rate of deformation in the specimen material is measured in terms of strain rate, i.e., strain/time, i.e., $\dot{\varepsilon} = \frac{\varepsilon}{t}$, whose units are 1/s or s^{-1}. For example, in Lab 3, the tension test on a steel specimen was conducted over a period of 1 m 40 s (total 100 s) and the total strain accumulated in the gage section was around 40%. Thus, the strain rate in this quasistatic test is $0.4/100 = 0.004/s$ or 4×10^{-3} s^{-1}. However, if the same steel is used in an automobile, in a collision event, the impact occurs in few hundred microseconds or few milliseconds. It is a dynamic event. Assuming the impact occurred in 100 ms and the failure strain was 40%, the strain rate is $= 0.4/(100 \times 10^{-3}) = 4/s$ or 4 s^{-1}, thousand times more than the strain rate achieved in a quasistatic conditions! If an engineer wants to optimally design an automobile for crash worthiness, the changes in the stress–strain behavior under dynamic loads must be carefully incorporated into the design process.

The discussion in the previous section on stress wave propagation characteristics using a short striker and two long rods can be used to develop dynamic stress–strain response of materials. This equipment is called Split Hopkinson Pressure Bar (SHPB) or Kolsky Bar. A schematic of the bar and the relevant equations are shown in Fig. 9.8. The two long bars with strain gages bonded on them are called the incident bar and transmission bar. The specimen material, whose dynamic stress–strain response is to be determined, is placed between the two bars. The incident bar is impacted by the striker bar (launched from a gas gun or propelled by other means). The impact generates a compressive stress wave in the incident bar. The wave travels along the length of the bar and reaches the bar-specimen interface. Due to the impedance mismatch between the bar material and the specimen, a portion of the stress wave passes through the specimen and into the transmission bar, and the rest is reflected back into the incident bar. During this process, the specimen is compressed. The strain gages bonded at mid-length of each bar capture the specimen response as the stress waves pass by. Typical strain gage signals measured on a SHPB while testing a metallic specimen are shown in Fig. 9.9. By measuring the incident, transmitted and reflected wave amplitudes and conducting 1D wave analysis, we can derive the following equations for stress in the specimen and strain in the specimen as [1]:

$$\sigma_{sp} = \frac{A_{bar} E_{bar}}{A_{sp}} \varepsilon_{tr}(t) \tag{9.4}$$

$$\varepsilon_{sp} = \frac{-2C_L}{l_{sp}} \int_0^t \varepsilon_{ref}(t) \tag{9.5}$$

where A_{bar} and E_{bar} are area of cross-section and Young's modulus of the bars, A_{sp} and l_{sp} are the area of cross-section and length of the specimen, ε_{tr} and ε_{ref} are transmitted strain and reflected strain measurements as a function of time (measured by the strain gages on transmitted and incident bars, respectively).

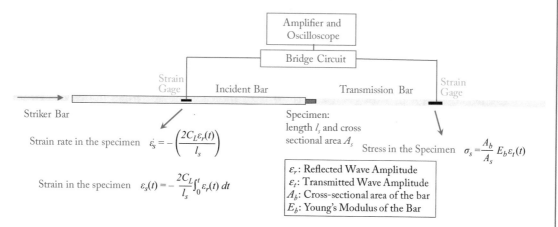

Figure 9.8: Schematic of the split Hopkinson pressure bar and the relevant equations for obtaining stress–strain response of a specimen material.

From the above equations, it is seen that the transmitted wave amplitude is a measure of specimen material's resistance (i.e., stress required to deform the material to finite strain) and the

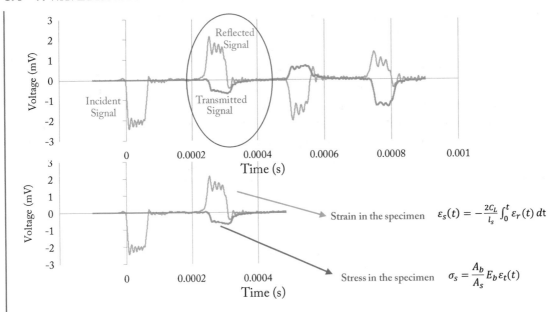

Figure 9.9: Wave signals measured by the strain gages on incident and transmission bars when the specimen is sandwiched between the bars.

reflected wave amplitude is a measure of deformation (strain rate) in the specimen. The strain rate in the specimen during deformation is given by

$$\dot{\varepsilon}_{sp} = \frac{-2C_L}{l_{sp}} \varepsilon_{ref}(t) \tag{9.6}$$

The negative sign in Eqs. (9.5) and (9.6) is to indicate the specimen undergoes compression. C_L is the longitudinal wave velocity in the bars. Note that integration of strain rate equation given in (9.6) is the strain (Eq. (9.5)). Thus, the dynamic stress–strain response of a material can be determined using the SHPB at a higher strain rate. For some commonly used materials (metals), the stress–strain response at quasistatic (10^{-3}/s) and dynamic (10^{3}/s) strain rates is provided in Fig. 9.10 [2]. Note that some metals exhibit rate-sensitive response, i.e., their stress–strain behavior is different under dynamic loads compared to quasistatic loads, and others are rate-insensitive. In general, face centered cubic metals (e.g., aluminum and copper) are rate-insensitive and metals with body centered cubic (e.g., iron or steel) and hexagonal (e.g., titanium) crystal structure are rate-sensitive. The differences in the stress–strain response can manifest in the form of increase in yield stress at higher rates and differences in work-hardening rate. In the case of ceramics or brittle materials which exhibit only elastic deformation, the failure strength increases significantly under dynamic loads, as illustrated schematically in Fig. 9.11.

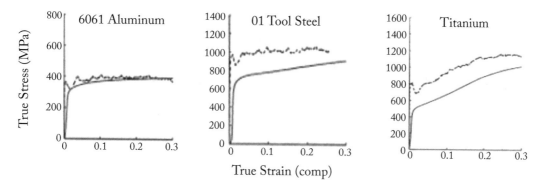

Figure 9.10: Stress–strain response of materials at quasistatic (10^{-3}/s) and dynamic (10^3/s) strain rates. Note that some materials are strain rate insensitive (e.g., aluminum) and others are strain rate sensitive (e.g., steel and aluminum). The dashed line is obtained at quasistatic strain rate and solid line is obtained dynamic conditions.

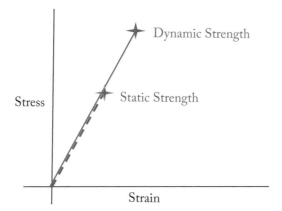

Figure 9.11: Schematic of stress–strain response of brittle materials at quasistatic (10^{-3}/s) and dynamic (10^3/s) strain rates. Note the increase in strength under dynamic strain rate compared to quasistatic strain rate.

Clearly, from the above discussion, we notice that the stress–strain response of a material is not unique. It changes with strain rate (also temperature). Although not discussed in this class, it is important to realize that the stress–strain response of a material also changes with strain (prior cold work), temperature, microstructure (grain size, defects, impurities, etc.), and many other factors.

REFERENCES

[1] M. A. Meyers, *Dynamic Behavior of Materials*, Wiley & Sons Inc., 2007. 195

[2] G. Subhash, *Dynamic Indentation Testing*, ASM Handbook®, Mechanical Testing and Evaluation, V. 8, ASM International, Materials Park, OH, pp. 519–529, 2000. 196

PART B: EXPERIMENT

9.9 WAVE PROPAGATION AND HIGH STRAIN RATE MATERIAL BEHAVIOR

9.9.1 OBJECTIVES

1. Investigate the 1D stress wave propagation in a slender rod.

2. Investigate the stress wave transmission to another rod of equal impedance and its reflection at a free boundary.

3. Investigate the stress–strain response of a ductile and a brittle material at high strain rates.

9.9.2 EQUIPMENT AND RESOURCES NEEDED

- Two long rods (incident and transmission bars), each of 1,200 mm length and 18 mm diameter and aligned on a steel beam and supports, as shown in Fig. 9.12. The rods are bonded with strain gages at the mid length. A striker bar of same diameter and 300 mm length, also aligned with the long rods. This arrangement with three bars is called split Hopkinson pressure bar (SHPB).

- Digital oscilloscope, Rigol DS4012, or similar. (Note that we are using a digital oscilloscope to record data because most laptop computers are not fast enough to capture the data at high speed.)

- Cylindrical specimens of dimension 4 mm diameter × 6 mm length of a metal and a ceramic.

- Strain gage amplifier, typical part: Tacuna Systems EMBSGB200_2_3 configured for 1/4-bridge. Two amplifiers are required because two strain gages are bonded in 1/2-bridge configuration on diametrically opposite side of each rod.

- A small rod on which two dummy gages are bonded to complete the W-bridge. By bonding these strain gages on a similar rod material we achieve thermal compensation.

- A mallet to propel and accelerate the striker bar.

9.9.3 EXPERIMENTAL TASK

Using the equipment given in the laboratory, perform the following tasks.

1. Measure the dimensions of the SHPB bars. Measure the location of strain gage with respect to one end of the bar. Connect the strain gages on the incident and transmission bars to W-bridge, power supply, amplifier, and oscilloscope.

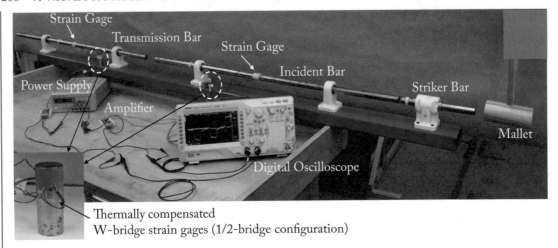

Figure 9.12: Split Hopkinson pressure bar with strain gages and an oscilloscope. Two strain gages are bonded on each bar (1/2-bridge configuration) and the other two are bonded on a smaller bar (shown in the inset) to provide thermal compensation and complete the W-bridge assembly.

2. Set the digital oscilloscope to capture the waves when they arrive at the strain gages. Set right time window to allow capture of at least four wave reflections in the incident bar. This window duration can be estimated by knowing the length of the incident bar and the wave velocity in the bar.

3. Let the incident and transmission bars be NOT in contact initially. Move the striker bar 2–3 cm away from the incident bar. Hit the striker bar briskly with a mallet so as to propel the striker toward the incident bar. Ensure that the mallet is not in contact with the striker by the time it reaches the incident bar. Upon hitting the striker, briskly remove the mallet. Capture the waves on an oscilloscope and observe the wave travel back and forth in the incident bar.

4. Bring the transmission bar in contact with the incident bar. Create the stress wave in the incident bar by hitting the striker bar with a mallet. Observe the wave propagation in incident and transmission bars.

5. Measure the dimensions of the specimens to be tested. Place a ductile specimen in between the incident and transmission bars and repeat step 2 to deform the specimen. Capture the wave signals.

6. Repeat step 5 with a brittle specimen.

9.9.4   ISSUES TO BE DISCUSSED IN THE LAB REPORT

1. Measure the duration of the stress pulse and calculate the wave velocity based on striker bar length.

2. Calculate the wave velocity based on the round trip travel time of the stress wave in the incident bar.

3. Compare the above two values with the theoretical formulation based on material Young's modulus and density. Discuss any discrepancies.

4. Using the wave signals generated in your experiment, discuss the wave propagation phenomena when a single bar is used and when two bars in contact are used. Also discuss reflection at the free-end.

5. In the context of experiments you have performed, discuss the wave behavior when a compression wave arrives at a junction and a tension wave arrives at a junction between two bars.

6. Using the wave signals you have obtained in your experiment, determine the stress–strain curve for the materials tested.

7. Calculate strain rate at which the specimens were deformed. Calculate strain rate based on theory of SHPB and also by measuring the specimen dimension before and after the test and the duration of stress pulse.

9.9.5   EQUIPMENT REQUIREMENTS AND SOURCING

- SHPB with strain gages on incident and transmission bars each of 18 mm diameter and 1,500 mm length. Striker bar of same diameter and 300 mm length. Gages in 1/2-bridge with 1/2-bridge compensating dummies on the same material as bars.

- Digital oscilloscope, Rigol DS4012 or similar.

- Strain gage amplifier, typical part: Tacuna Systems EMBSGB200_2_3 configured for 1/4-bridge

- Compressive test specimens: Marble cubes ~ 10 mm on a side. Brass washers 8 mm diameter, 4 mm thick.

Authors' Biographies

GHATU SUBHASH

Professor Ghatu Subhash is Newton C. Ebaugh Professor of Mechanical and Aerospace Engineering at the University of Florida, Gainesville, FL. He obtained his M.S. and Ph.D. from the University of California San Diego and conducted post-doctoral research at California Institute of Technology, Pasadena, CA. He joined Michigan Technological University in 1993 and then moved to the University of Florida in 2007. He has received numerous honors for excellence in teaching, research, and professional service, including the Frocht Award from Society for Experimental Mechanics (SEM), Best Paper – *Journal of Engineering Materials and Technology*; Significant Contribution Award American Nuclear Society – Materials Science and Technology Division; Fellow of SEM, Fellow of American Society of Mechanical Engineers (ASME); Technology Innovator Award, University of Florida; ASME Student Section Advisor Award; Society of Automotive Engineers (SAE) Ralph R. Teetor Educational Award; American Society for Engineering Education (ASEE) Outstanding New Mechanics Educator award; A Distinguished Faculty Member by the Michigan Association of Governing Boards of State Universities; and Commendation Letters from the Governor of Michigan. He has served as an Associate Editor for the *Journal of the American Ceramic Society, Mechanics of Materials Experimental Mechanics, Journal of Engineering Materials and Technology*, and *Journal of Dynamic Behavior of Materials*. He has graduated 30 Ph.D. students and is currently advising 6 Ph.D. students in various fields related to multiaxial behavior and microstructural characterization of advanced materials, ultrahard ceramics, composites, and bio materials. He has authored more than 250 scientific articles and 3 patents. He has given numerous invited lectures and seminars at various conferences and universities. His invention of fluid-filled energy absorbing cushions has appeared on several TV networks, radio stations (including NPR), and in articles by Reuters, ASEE Morning Bell, and many newspapers. Most recently, he appeared in a PBS documentary "Secrets of Spanish Florida", aired nationwide in December 2017 while discussing the impact response of Coquina, the material with which the oldest fort in the U.S., Castillo de San Marcos, in St. Augustine, Florida was built.

SHANNON RIDGEWAY

Shannon Ridgeway is an Instructor in the Department of Mechanical and Aerospace Engineering at the University of Florida. He received his B.S. in Mechanical Engineering from the Florida Institute of Technology in 1987. He was employed as an engineer designing custom contact molded fiberglass reinforced products for Satsuma Inc. following graduation. He then went on to continue his studies and received his M.S. in Mechanical Engineering from the University of Florida in 1995.

Mr. Ridgeway has been involved in research in the areas of design, mechanics of materials, spatial mechanisms, robotics, controls systems, and autonomous navigation for over 20 years. He participated in the Center for Intelligent Machines and Robotics (CIMAR) entries into the DARPA Grand Challenges and Urban Challenge, primarily working on hardware integration and system design. Current activities include instructing the Mechanics of Materials Undergraduate Laboratory and the Dynamic Systems and Controls Undergraduate Laboratory for Mechanical and Aerospace Engineering, at the University of Florida. Current research interests focus on ground and surface vehicle autonomy.

Printed in the United States
by Baker & Taylor Publisher Services